Pirate Philosophy

Closer: Performance, Technologies, Phenomenology, Susan Kozel, 2007
Video: The Reflexive Medium, Yvonne Spielmann, 2007
Software Studies: A Lexicon, Matthew Fuller, 2008
Tactical Biopolitics: Art, Activism, and Technoscience, edited by Beatriz da Costa and Kavita Philip, 2008
White Heat and Cold Logic: British Computer Art 1960–1980, edited by Paul Brown, Charlie Gere, Nicholas Lambert, and Catherine Mason, 2008
Rethinking Curating: Art after New Media, Beryl Graham and Sarah Cook, 2010
Green Light: Toward an Art of Evolution, George Gessert, 2010
Enfoldment and Infinity: An Islamic Genealogy of New Media Art, Laura U. Marks, 2010
Synthetics: Aspects of Art & Technology in Australia, 1956–1975, Stephen Jones, 2011
Hybrid Cultures: Japanese Media Arts in Dialogue with the West, Yvonne Spielmann, 2012
Walking and Mapping: Artists as Cartographers, Karen O'Rourke, 2013
The Fourth Dimension and Non-Euclidean Geometry in Modern Art, revised edition, Linda Dalrymple Henderson, 2013
Illusions in Motion: Media Archaeology of the Moving Panorama and Related Spectacles, Erkki Huhtamo, 2013
Relive: Media Art Histories, edited by Sean Cubitt and Paul Thomas, 2013
Re-collection: Art, New Media, and Social Memory, Richard Rinehart and Jon Ippolito, 2014
Biopolitical Screens: Image, Power, and the Neoliberal Brain, Pasi Väliaho, 2014
The Practice of Light: A Genealogy of Visual Technologies from Prints to Pixels, Sean Cubitt, 2014
The Tone of Our Times: Sound, Sense, Economy, and Ecology, Frances Dyson, 2014
The Experience Machine: Stan VanDerBeek's Movie-Drome and Expanded Cinema, Gloria Sutton, 2014
Hanan al-Cinema: Affections for the Moving Image, Laura U. Marks, 2015
Writing and Unwriting (Media) Art History: Erkki Kurenniemi in 2048, edited by Joasia Krysa and Jussi Parikka, 2015
Control: Digitality as Cultural Logic, Seb Franklin, 2015
New Tendencies: Art at the Threshold of the Information Revolution (1961–1978), Armin Medosch, 2016
Screen Ecologies: Art, Media, and the Environment in the Asia-Pacific Region, Larissa Hjorth, Sarah Pink, Kristen Sharp, and Linda Williams, 2016
Pirate Philosophy: For a Digital Posthumanities, Gary Hall, 2016

See http://mitpress.mit.edu for a complete list of titles in this series.

Pirate Philosophy

For a Digital Posthumanties

Gary Hall

The MIT Press
Cambridge, Massachusetts
London, England

This book was set in Stone Sans and Stone Serif by Toppan Best-set Premedia Limited. Printed and bound in the United States of America.

Library of Congress Cataloging-in-Publication Data

Names: Hall, Gary, 1962– author.
Title: Pirate philosophy for a digital posthumanities / Gary Hall.
Description: Cambridge, MA: The MIT Press, 2016. | Series: Leonardo book series | Includes bibliographical references and index.
Identifiers: LCCN 2015039700 | ISBN 9780262034401 (hardcover : alk. paper)
Subjects: LCSH: Humanities—Technological innovations. | Humanities—Research. | Communication in learning and scholarship—Technological innovations. | Scholarly publishing—Technological innovations. | Open access publishing.
Classification: LCC AZ195 .H35 2016 | DDC 001.3—dc23 LC record available at http://lccn.loc.gov/2015039700

10 9 8 7 6 5 4 3 2 1

This one is for Mark Poster

Contents

Series Foreword

Leonardo/International Society for the Arts, Sciences, and Technology (ISAST)

Leonardo, the International Society for the Arts, Sciences, and Technology, and the affiliated French organization Association Leonardo have some very simple goals:

1. To advocate, document, and make known the work of artists, researchers, and scholars developing the new ways that the contemporary arts interact with science, technology, and society.
2. To create a forum and meeting places where artists, scientists, and engineers can meet, exchange ideas, and, when appropriate, collaborate.
3. To contribute, through the interaction of the arts and sciences, to the creation of the new culture that will be needed to transition to a sustainable planetary society.

When the journal *Leonardo* was started some forty-five years ago, these creative disciplines existed in segregated institutional and social networks, a situation dramatized at that time by the "two cultures" debates initiated by C. P. Snow. Today we live in a different time of cross-disciplinary ferment, collaboration, and intellectual confrontation enabled by new hybrid organizations, new funding sponsors, and the shared tools of computers and the Internet. Above all, new generations of artist-researchers and researcher-artists are now at work individually and collaboratively bridging the art, science, and technology disciplines. For some of the hard problems in our society, we have no choice but to find new ways to couple the arts and sciences. Perhaps in our lifetime we will see the emergence of "new Leonardos," hybrid creative individuals or teams that will not only develop a meaningful art for our times but also drive new agendas in science and stimulate technological innovation that addresses today's human needs.

For more information on the activities of the Leonardo organizations and networks, please visit our websites at http://www.leonardo.info/ and http://www.olats.org.

Roger F. Malina
Executive Editor, Leonardo Publications

Acknowledgments

Many of the ideas contained in this book were developed while I was visiting fellow in the Centre for Research in the Arts, Humanities and Social Sciences (CRASSH) at the University of Cambridge and visiting professor at the Hybrid Publishing Lab—Leuphana Inkubator, Leuphana University, Germany. I thank my friends and colleagues at both institutions for their intellectual hospitality and generosity. I also presented material that is in this book at the European University Institute, Florence; University of St Andrews; University of Leuven; University of Kent; London School of Economics; Whitchapel Gallery, London; Aarhus University; University of Sussex; Universidad de las Americas, Puebla; University of Beira Interior, Covilhã; Eastern Mediterranean University, North Cyprus; Goldsmiths, University of London; University of Birmingham; University of East London; Reina Sofia Museum, Madrid; and Amsterdam Central Library and the Royal Library in the Hague. My thanks to all those who provided me with the gift of feedback, comments, questions, and suggestions on these occasions.

Special thanks are due to Janneke Adema, Mark Amerika, Clare Birchall, Dave Boothroyd, Gabriela Méndez Cota, Alberto López Cuenca, Jeremy Gilbert, Sigi Jöttkandt, Sara Kember, Benjamín Mayer Foulkes, Tara McPherson, David Ottina, John Carlos Rowe, Nina Sellars, Doug Sery, Marq Smith, and Stelarc; my colleagues at Coventry; and MIT's anonymous reviewers. Extra special thanks are due to Joanna Zylinska, as always.

An earlier version of chapter 2 appeared as "Towards a Post-Digital Humanities: Cultural Analytics and the Computational Turn in the Digital Humanities," *American Literature* 85, no. 4 (December 2013), special issue editors: Tara McPherson, Patrick Jagoda, and Wendy H. K. Chun; chapter 3 as "#MySubjectivation," *New Formations*, no. 79 (Autumn 2013); chapter 5 as "Pirate Radical Philosophy," *Radical Philosophy: A Journal of Socialist and*

Feminist Philosophy 173 (May/June 2012); and chapter 6 as "The Unbound Book: Scholarly Publishing in the Age of the Infinite Archive," *Journal of Visual Culture* 15 no. 12 (2013). I have revised and extended all of this work for the purposes of this book, as I have the "Copyfights" section of chapter 1, an earlier version of which appeared as 'Copyfight' Critical Keywords for the Digital Humanities (Lüneburg: Centre for Digital Cultures, Leuphana University, 2014).

Preface

We find ourselves in a *time of riots* wherein a rebirth of History, as opposed to the pure and simple repetition of the worst, is signalled and takes shape.[1]—Alain Badiou

Since the financial crash of 2008, much has been written about the "crisis of capitalism" and the associated series of postcrash political events that are seen as having begun with the Tunisian revolution of 2010: the Arab Spring, the 2011 "August riots" in England, and Occupy Wall Street, together with the movement of the European squares that eventually led to the election of the radical left Syriza party in Greece and rise to prominence of another left-wing party, Podemos, in Spain. Yet to what extent does our contemporary sociopolitical situation also pose a challenge to those of us who work and study in the university? How can we act not so much for or with the antiausterity and student protesters, "graduates without a future," and "remainder of capital," demonstrating alongside them, accepting invitations to speak to them and write about them and so on, but rather in terms of them, thus refusing to submit critical thought to "existing political discourses and the formulation of political needs those discourses articulate," and so "'defusing the trap of the event'"?[2] Does the struggle against the neoliberal corporatization of higher education not require *us* to have the courage to transform radically the material practices and social relations of our lives and labor?

These questions form the starting point for this book's engagement with a range of theorists and philosophers, operating in some of the most exciting and cutting-edge areas of the humanities today. They include Lev Manovich (the digital humanities), Rosi Braidotti (new materialism), Bernard Stiegler (posthumanism), and Graham Harman (object-oriented ontology). Drawing critically on phenomena such as the peer-to-peer file-sharing and the anticopyright pro-piracy movements, *Pirate Philosophy* explores how we can produce not just new ways of thinking about the world, which is what theorists and philosophers have traditionally aspired to do, but new ways of actually being theorists and philosophers in this "time of riots."

The book's opening chapter sets the scene with an account of the politics of online sharing in relation to the struggles against the current intellectual property regime associated with Anonymous, LulzSec, Aaron Swartz, and the "academic spring" of 2012. It discusses Creative Commons; the open access, open source, and free software movements; and the difficulty of forging a common, oppositional horizon given that these struggles and movements do not share a common idea of the Commons. In the chapters that follow, *Pirate Philosophy* proceeds to ask how, when it comes to our own scholarly ways of creating, performing, and sharing knowledge and research, we can operate in a manner that is different not just from the neoliberal model of the entrepreneurial academic associated with corporate social networks such as Facebook and LinkedIn, but also from the traditional liberal humanist model that comes replete with clichéd, ready-made (some would even say cowardly) ideas of proprietorial authorship, the book, originality, fixity, and the finished object.

Of course, many theorists are challenging the dictatorship of the human with an emphasis on the nonhuman, the posthuman, and the postanthropocentric, along with the crisis of life itself that is expressed by the Anthropocene. Yet such "posttheory theories" continue to remain intricately bound up with humanism and the human in the very performance of their attempt to think beyond them due to the approaches they have adopted in response to the question of the politics of copying, distributing, selling, and (re)using theory. This is to some extent inevitable given the lack of antihumanist alternatives to publishing either on a "copyright ... all rights reserved" or open access and Creative Commons basis that are institutionally and professionally recognized. Nevertheless, *Pirate Philosophy* endeavors to move the analysis of the human and nonhuman on by raising a question that is also an exhortation: How, as theorists and philosophers, can we act differently—to the point where we begin to take on and assume some of the implications of the challenge that is offered by theory and philosophy to fundamental humanities concepts such as the human, the subject, the author, copyright, community, and the common, for the ways in which we live, work, and think? How, in other words, can we act as something like pirate philosophers in the sense of the term's etymological origins with the ancient Greeks, where the pirate is someone who tries, tests, teases, and troubles, as well as attacks? Might doing so be one way for us to try out and put to the test new economic, legal, and political models for the creation, publication, and circulation of knowledge and ideas, models that are more appropriate for our postcrash sociopolitical situation?

1 THE COMMONS AND COMMUNITY

How We Remain Modern

Pirate ... from the Latin *pirata* (*-ae;* pirate) ... transliteration of the Greek *piratis* (pirate) from the verb *pirao* (make an attempt, try, test, get experience, endeavour, attack ...).
In modern Greek ... *piragma*: teasing ... *pirazo*: tease, give trouble[1]

Copyfights

Not so long ago, large-scale political protest seemed to be a thing of the past. It was as if this form of political activism had more or less come to an end with the anticapitalist globalization movements of the pre-9/11 world; and if not then, certainly with the antiwar marches of 2003 and their failure to prevent the subsequent invasion of Iraq. The years following the financial crash of 2008, however, have seen the "Arab renaissance," the worldwide Occupy movements, and antiausterity and student protests usher in a new age of mass mobilization. We now live in an era characterized by a widespread rejection of the principle of political representation and individual fame and by attempts to develop nonhierarchical forms of political organization and coordination instead.[2] And as the activities of international hacktivist networks such as Anonymous and LulzSec bear witness, similar characteristics are a feature of many of the related struggles around copyright, intellectual property, and Internet piracy.

To be sure, some conflicts with the current Euro American intellectual property regime have been won. The service blackout coordinated by Wikipedia and others in January 2012 resulted in the Stop Online Piracy Act (SOPA) and Protect IP Act (PIPA) bills being postponed in the United States. The Academic Spring of the same year, in which over 12,000 academics signed a public petition protesting the business practices of the publisher Elsevier, reported to make 725 million euros in annual profits on its journals alone, had a similar effect on the Research Works Act. Still, the overall victors in the copyright wars are undoubtedly the multinational conglomerates of the cultural industries that, with the backing of governments worldwide, continue to control (albeit in a fashion that is not without interruption or failure) the production, distribution, and marketing of the majority of our knowledge and information. Witness the federal charges brought by the US Department of Justice in July 2011 against the self-declared open access guerrilla Aaron Swartz for his large-scale unauthorized downloading of files from the JSTOR academic database. Swartz, a founder of the online activist group Demand Progress, which launched the campaign against SOPA and PIPA, was threatened with a thirty-five-year prison sentence and a fine of $1million. He committed suicide in January 2013 before his case came to trial.

Indeed, for many political activists and theorists, the situation over the right to access, copy, distribute, sell, and (re)use artistic, literary, cultural, and academic research works and other materials is, if anything, getting worse. They cite as evidence a profound shift that is taking place in the digital world. It is a shift toward the closed, centralized systems of mobile media and the cloud, as represented by the nonconfigurable iDevices and single-purpose apps of Apple that are designed to optimize the distribution of media simply as commodities to be purchased and consumed (after the launch of the iPhone 5, Apple became the most valuable company of all time in terms of market capitalization) and away from the open, distributed networks and physical infrastructure of the Web that allows users to understand how such digital products are made and to copy, share, change, update, improve, and reimagine them continually. Coupled to the fast-emerging online media monopolies of a small number of extremely rich and powerful international corporations, including Apple, Amazon, Facebook, and Google, and viewed in the light of the 2013 Edward Snowden disclosures especially, it is a transition that has led some to predict the death of the open Web.[3]

Radical theorists of politics and the media are thus confronted by some key questions. How might we turn from intellectual property laws and infrastructure designed for the benefit of "the 1 percent" to find ways of openly sharing art, education, knowledge, and culture, while at the same time ensuring creative workers are adequately and justly compensated for their labor? Is this primarily a cultural issue? Or does it require the development of new laws, new forms of political organization, new economies—even new ways of organizing postindustrial society? And if the latter is the case, then how could something of this sort be achieved given that, as Felix

Stalder points out, many of the oppositional organizational movements associated with struggles over copyright, intellectual property, and Internet piracy find it extremely difficult to "engage in the institutional world in any other than destructive ways"? Anonymous, for example, "cannot, and does not aim at building alternative institutions," Stalder emphasizes. He does, however, see this informal grouping as being capable of "contributing to the forging of a common, oppositional horizon that could make it easier to coordinate further action."[4]

This state of events raises an additional question about the possible forms the forging of such a common, oppositional horizon might take. It is not difficult to envisage any coordinated further action in the future endeavoring to include Creative Commons and the open access, free software, peer-to-peer file sharing, and anticopyright pro-piracy movements on the basis they all offer a challenge of one kind or another to the current intellectual property regime. Yet how much do such initiatives actually have in common? How significant is it that they do not even share a common idea of the Commons?

Creative Commons (CC) is a nonprofit organization that offers a range of easy-to-use copyright licenses that authors and artists can choose from in order to grant others permission to share their work and use it creatively. Rather than the default copyright position of all rights reserved, CC licenses range from some rights reserved to a public domain CC-0 license that waives all rights. Creative Commons thus provides a means of protecting the rights of creators from the extremes of intellectual property law, including the length to which copyright has been extended as a result of lobbying from companies such as Disney.

Open access, meanwhile, is concerned with making academic research openly available online. Many texts published on an open access basis are covered by a CC license that permits them to be openly read, copied, and distributed but not built on, developed, altered, and improved by others in the way that free and open source software is. A substantial number of open access initiatives have undergone a change in licensing policy in recent years, however. More and more have adopted a Creative Commons CC-BY license that insists only on author attribution, thus giving others permission to copy and reuse texts and make derivative works from them. (The announcement in late 2014 by the Bill and Melinda Gates Foundation that from 2017 on, all research papers generated as a result of their funding must be made available open access with a CC-BY license immediately on publication is merely one of the latest initiatives of this kind.) To a certain extent, this change has been motivated by a concern to grant users open access not merely to the research but to the associated data too. This includes the right to mine texts and data, as some permission barriers can block text and data mining. (That said, it is worth pointing out that a CC-BY license is not enough on its own to ensure text and data mining are possible. A PDF, for example, cannot be easily mined for data regardless of whether it is published on a CC-BY basis.) There are some open access advocates, however, who view this shift in policy to CC-BY licensing as going too far. They argue that opening up access to research has to be the priority, and that any insistence academics do so on a basis that allows others to modify work as well will succeed only in alienating the majority of the research community from publishing open access in the first place (e.g., because the support for the CC-BY license risks being perceived as a plagiarist's charter).

Yet both of these positions—for and against the Creative Commons CC-BY license—are criticized by many theorists in certain more politically engaged areas of new media studies, software studies, and cultural studies. In fact, they disagree with the very notion of the Commons as it is used and understood by Creative Commons. For them, Creative Commons is more concerned with preserving the rights of copyright owners than with granting rights to users. Creative Commons is also extremely liberal and individualistic: rather than endorsing a collective agreement, policy, or philosophy, it provides a range of licenses from which authors can individually choose. (And this is so even in the case of the public domain CC-0 license that waives all rights.) Contrary to the way Creative Commons is frequently portrayed, then, it is not advocating a common stock of nonprivately owned (creative) works that everyone jointly manages, shares, and is free to access and use on the same basis at all, which is how the Commons is often understood.[5] Instead, Creative Commons presumes everything created by an author or artist is that person's property. What Creative Commons offers is therefore not so much a fundamental critique of intellectual property (IP) law or a challenge to it as merely a reform of it.[6] Creative Commons is simply helping the law adapt to the new conditions created by digital culture by supporting a smarter, more open, and flexible model of individual ownership.

In this respect, the emphasis of Creative Commons on the rights of copyright owners has a clear strategic purpose, in that it speaks to what Andrew Ross describes as "the thwarted class fraction of high-skilled and self-directed individuals in the creative and knowledge sectors"—academics, artists, designers, musicians, writers, software developers and so on—"whose entrepreneurial prospects are increasingly blocked by corporate monopolies."[7] Exponents of this understanding of copyright have thus been able to form a "coalition of experts with the legal access and resources" to mount a powerful campaign that frequently overshadows often more interesting and radical approaches (161). This explains why the issue is so closely associated with Lawrence Lessig, James Boyle, and Cory Doctorow, and their reformist lobbying for better IP law (i.e., IP law that does not put business, competition, and innovation at risk), rather than no IP law or a radically different IP law. It also clarifies why CC licenses are so widely used in open access. The result is that this aspect of the debate over free culture risks being, in Ross's words, "simply an elite copyfight between capital-owner monopolists and the labor aristocracy of the digitariat (a dominated fraction of the dominant class, as Pierre Bourdieu once described intellectuals) struggling to preserve and extend their high-skill interests" (169).

Many in the free software community, including Richard Stallman, founder of the Free Software Foundation and inventor of the general public license, the most common free software copyright license, lobby for what is called copyleft instead. Like Creative Commons, this approach still entails a use of IP law, only it is designed to serve the opposite ends to those to which such a license is usually put. Rather than supporting the ownership of private property, copyleft defends the freedom of everyone to copy, distribute, develop, and improve software or any other work covered by such a license. The only permission barrier is that which upholds this right by insisting all such copies and derivatives must be shared under the same terms and conditions, thereby ensuring that the freedom of everyone to do likewise continues into the future and denying anyone a competitive advantage.

The free software community likes to position itself as a social movement and as being more ethically and politically engaged than those who argue for Creative Commons and open source. Whereas the former encourages collaborative working (even though that collaboration is regarded as more of a means to an end than an end in itself), Creative Commons is held as being quite individualistic—not just in the way a particular CC license is applied but also in how CC-licensed works tend to be used. Similarly, in their concern to determine the best way, practically, to develop a (potentially monetizable) product in an open manner and not alienate the world of corporate capital by using terms such as *free* that could all too easily be ascribed to a radical left approach to property, those associated with open source are perceived from within the movement for free software as adhering to the logic of the market in too pragmatic a fashion. Yet many activists and theorists question just how left politically copyleft actually is. Free software made available on this basis is not necessarily anticommercial or anticapitalist: in Stallman's words, "think of free speech, not free beer." So as long as they are covered by the same license, there is nothing to prevent a corporation from selling copies of the software it has developed from the original source code. (To provide an example, while making the source code available for free, thus respecting the terms of the copyleft license, it could sell the executable application a user needs to actually run the software and which they may have neither the time nor the skills to produce for themselves.) Indeed, for some, both free *and* open source software (F/OSS), as forms of collaborative peer production, have been subsumed into a mode of economic development that is dominated by large corporate actors. Companies such as IBM, Microsoft, Google, and others see in F/OSS a means of acquiring software and the related research, development, and customer support at comparatively little expense, without necessarily offering all that much in return. In doing so, these monopolistic multinationals go against the ethos of reciprocity and for-benefit (rather than for-profit) that is often associated with such collaborative forms of peer production. They also help to maintain a situation in which many F/OSS communities are neither autonomous nor self-sustainable, as most of those contributing to them are by necessity laboring for capital in some shape or form. Moreover, the IP that copyleft aims at is, as David Golumbia emphasizes, rarely that which is of most value to capital: "corporate secrets, scientific IP in private hands, etc." Capital already has other means at its disposal for protecting that. Instead, copyleft is directed at the IP produced by those who tend to be in a relatively precarious position already: academics, artists, designers, and software developers, for example.[8]

There is also the problem that, in contrast to Creative Commons, the philosophy of free software cannot be easily applied to other areas of culture to create a larger Commons, for the simple reason that this philosophy does not scale. It is an aspect of the situation that is encapsulated by the software developer and founder of the Telekommunisten Collective, Dmytri Kleiner, as follows:

> Companies for whom software is a necessary capital input are happy to support [the production and development of] free software, because doing so is most often more beneficial to them than either paying for proprietary software, or developing their own systems from scratch. They make their profit from the goods

and services which they produce, not from the software they employ in their production.

Cultural Works, especially popular ones, such as book[s], movies, music, etc, are not usually producer's goods. In a capitalism economy these are generally Consumer's goods, and thus the publishers of such works must capture profit on their circulation.

Thus capital will not finance free culture in the same way it has financed free software.[9]

Rather than preventing access to cultural works and source code from being restricted, those on the radical left tend to be more concerned with developing a free, common culture jointly managed and shared by all, and with promoting the equal and just distribution of wealth among the workers who produce and maintain it. To this end, Kleiner insists copyleft must be transformed into copyfarleft in which creative workers themselves own the means of production and only prevent use of their works that is *not* based in the Commons. This last point is especially important with regard to how workers can be compensated for their labor in the context of a free common culture. It means creative workers can "earn remuneration by applying their own labor to mutual property," but those who exploit wage labor and private property in production cannot.[10]

For copyfarleft to be able to generate such a worker-controlled economy, however, and thus itself succeed in having an impact on anything even approaching a significant social and political scale, it would need to be part of a much larger economy of this kind—one capable of taking in not just the production of art, culture, and software but basic items such as food and housing. Since the prospect of such an economy emerging any time soon looks unlikely, Kleiner acknowledges that complete anticopyright, as a radical gesture that "refuses pragmatic compromises and seeks to abolish intellectual property in its entirety," has significant appeal for many (42). This is especially true of those in the peer-to-peer file- and text-sharing communities where distinctions between producer and consumer are difficult to maintain. Some anti-intellectual property advocates in the pro-piracy movement even argue against copyright and the use of licenses altogether, regarding them as remnants from a previous age and inappropriate for an era in which artistic, literary, cultural, and academic research works can be copied and shared at very little expense, without depriving the original "owners" of their versions due to the nonrivalrous nature of digital objects. Instead of Creative Commons, they argue for a "gray commons," *gray* being used to signal the legal ambiguity of much of its content (that it is not a black-and-white issue). The gray commons thus connects to the "pirate" desire to avoid the formation of the type of organizational centers and hierarchies of authority and leadership that would inevitably ensue if anyone (e.g., platform managers, administrators, curators) were to be placed in a position that requires them to make decisions about what the Commons should and should not include, be it pirated music, videos of beheadings, or nude photos of female film stars.[11]

The difficulty with the anticopyright stance is that it may be effective only from a position either outside the capitalist legal system or after its demise. Certainly when it comes to academic publishing, gestures of this kind risk playing into the hands of the neoliberal philosophy that states universities should carry out the basic research the private sector does not have the time, money, or inclination to conduct for itself, while nevertheless granting businesses easy access to that research and the associated data for commercial application and exploitation. (This is another explanation for the shift in licensing policy within the open access movement toward CC-BY: it is designed to serve the neoliberal goal of enabling that which is available publicly and for the common good to be enclosed by private interests and traded on the market.)

All of this serves as a neat illustration of a paradox that Roberto Esposito locates in the very idea of the common in his book *Communitas: The Origin and Destiny of Community*. The paradox concerns the way in which "the 'common' is defined exactly through its most obvious antonym: what is common is that which unites the ethnic, territorial, and spiritual property of every one of its members. They have in common what is most properly their own; they are the owners of what is common to them all."[12] And to be sure, the Commons is a place where the interests of a large number of diverse groups, movements, organizations, initiatives, and constituencies regarding the right to copy, sell, use, and share—including artists, activists, academics, educators, and programmers belonging to both wider "majoritarian" and counterpublics—come together but also exist in a state of tension and conflict and are in fact often demonstrably incompatible and incommensurable. This is not to suggest that a coordinated, oppositional community of artists, activists, programmers, and so on is impossible to achieve. As Jean-Luc Nancy writes, "To be with, to be together, and even to be 'united' is precisely not to be 'one.' Of communities that are at one with themselves, there are only dead ones."[13] It is merely to acknowledge that a certain amount of antagonism and dissensus is what makes both the common and community possible and that if we *do* want to forge a common, oppositional horizon that could make it simpler to organize any future action over the right to copy, distribute, sell, and reuse artistic, literary, cultural, and academic research works and other materials, then we need to think the nature of community, of being together and holding something in common, differently.

What makes Esposito's *Communitas* so interesting from this point of view is its attempt to provide us with one way to begin to do just this and think community and the common otherwise. Esposito begins by showing that in "all neo-Latin languages (though not only), 'common' (*commun, comun, kommun*) is what is *not proper*."[14] From this starting point he proceeds to develop a notion of the common and community that brings into question and decenters the unified, sovereign, proprietorial subject—a subject on which, as we have seen, the Creative Commons, open access, and free software movements all depend:

> The common is not characterized by what is proper but by what is improper, or even more drastically, by the other; by a voiding, be it partial or whole, of property into its negative; by removing what is properly one's own that invests and decenters the proprietary subject, forcing him to take leave of himself, to alter himself. In the community, subjects do not find a principle of identification nor an aseptic enclosure within which they can establish transparent communication or even a content to be communicated. They don't find anything else except that void, that distance, that extraneousness that constitutes them as being missing from themselves. (7)

Just as interesting, especially in the light of what follows, is the fact that one way of thinking about the central void of community for Esposito is in terms of the gift.

Postcrash Critical Theory

It is not just notions of the common, community, and the sovereign subject that such an analysis of intellectual property, copyright, Creative Commons, open access, and Internet piracy raises questions for. The university today is one of the few places where the imposition of neoliberalism and its emphasis on production, privatization, and the interests of the market is still being directly resisted, to some extent at least. (No doubt this is one reason that the police and state in England and elsewhere are greeting protests over the future direction of higher education with a surprising degree of violence.)[15] Nevertheless, if cultural forms such as music, film, and television are lagging behind the change in political mood post-2008[16]—continuing to be dominated, as Rhian Jones says of the music industry, by "careerist and commercial imperatives," "hostility towards the new, untried and perhaps unprofitable," "reliance on resuscitating previous forms at the expense of innovation," and "the social stratification of access"—the same can be said of those cultural forms associated with the production, publication, and distribution of academic knowledge and research.[17]

In large part this is due to the fact that publishing, certainly as far as the majority of academics in the humanities are concerned, is just something they are required to be involved in as part of their professional activity. They may think about politics in relation to culture and society or even other parts of the media (e.g., the BBC, Twitter, algorithmic regulation and surveillance). Yet unless they are involved in the movements for Creative Commons, open access, free software, peer-to-peer file sharing, or pro-piracy—and even then in many cases—they do not spend too much time reflecting on the politics of their own knowledge production, let alone trying to challenge or change it. Linearly written and organized, bound, and printed paper codex books and journal articles are objects scholars in the humanities are constantly writing, reading, browsing, carrying, holding, making notes in, borrowing from and returning to libraries, buying in bookshops, ordering from online retailers, storing on office shelves, displaying in their homes, sharing with friends, and exchanging as gifts. But as a general rule, they pay little heed to what it means to publish with a profit-maximizing transnational corporation rather than with a nonprofit publisher or on an "all rights reserved" rather than Creative Commons or copyfarleft basis. As a politico-institutional system of production and control, complete with its inherited disciplinary practices and protocols of status, advancement, recognition, and credibility, it has become so naturalized and accepted that unless something out of the ordinary happens in the publishing process, a campaign is mounted (e.g., the 2012 boycott of Elsevier), or a change of policy is imposed at the government, funding agency, or institutional level, it is not something academics are especially willing or able to devote much critical attention to. All too frequently, it is simply there for them as a fundamental part of their world and everyday practice, yet for the most part unremarked on, unnoticed, unthought.

This is a particular issue for those discourses in the humanities that it is actually quite "difficult to classify or to name univocally, for one of the things they share is precisely the radical questioning of all such univocity," but which, first in the United States, and later elsewhere, have regularly been placed under the heading of "theory."[18] Initially associated with the study of literature and contemporary French philosophy, theory has helped a diverse range of new critical thinking to emerge from gender and postcolonial studies through science and technology studies, to software studies and beyond. And one of the reasons theory continues to be so important is its ability to denaturalize and destabilize institutional and disciplinary formations, including those associated with theory itself. Of course, it is impossible to achieve a perfect awareness of the parameters within which our own professional forms, methods, and procedures of knowledge operate. Nevertheless, theory is committed to challenging and changing our ways of being and acting in the world. And as David Theo Goldberg points out in a text that (like this one) is very much concerned with the future or "afterlife" of the humanities, theory must therefore by necessity "take on the social and thought frames themselves, the structural conditions constituting the conditions of possibility" that order and shape our "established and inherited ways of being, thinking, and doing." Yet as he goes on to make clear, for all this,

> much that goes under the rubric of the "critical," in theory, as in thinking more generally, assumes the frame of the conventional and given, the taken for granted social arrangements we inhabit. In that sense they change little if anything of the structural arrangements, material, political or institutional social worlds we inhabit and they take themselves to be addressing. They engage, overwhelmingly, in thinking *within* the faded frame.[19]

Goldberg does not analyze the specific frames that constitute the conditions of possibility of critical theory in any great detail in "The Afterlife of the Humanities." (He prefers to provide a sketch of what the humanities could look like if they were dramatically reconceived and remade so as to be able to "speak to our times.")[20] Nowhere is theory's thinking within the faded frame more evident, however, than in the way it continues to be dominated by the print-on-paper codex book and journal article, together with many of the core humanities concepts that have been inherited with them (which are of course not natural, but the result of years of historical development). The latter include a number of those concepts I have already begun to raise questions for in the process of analyzing the politics of sharing and the Commons, such as the unified, sovereign, proprietorial subject; the individualized author; intellectual property; and copyright. But as we see over the course of *Pirate Philosophy*, they also include the signature, the proper noun or name, originality, the finished object, immutability or "fixity," the book, the canon, the discipline, tradition, even the human, along with the institutions that sustain and support them: the university, the library, the publishing house, and so on. (And it is worth emphasizing that this domination is the case in spite of the fact that theory also has the potential to help us to critically and creatively explore, experiment with, and reinvent such interrelated humanities concepts as we have already witnessed to a certain degree with the example of Esposito's analyses of the common, community, and the proprietorial subject.) Indeed, if Western philosophy has forgotten that its origins lie with technics, if it has "repressed technics as an object of thought," as Bernard Stiegler insists, then many theorists and philosophers can be said to have also forgotten and repressed the technologies by which their own work is not only produced, published, and distributed but also commodified and privatized (not to mention controlled, homogenized, and standardized) by for-profit companies operating as part of the cultural industries.[21] And, ironically enough, this includes Stiegler himself.

Nor is a lack of care and attention to the politics of their own knowledge production and postproduction confined merely to those instances when the communication technologies in question are those that historically have been most commonly employed by theorists and philosophers: the book and journal article published with a traditional print press. A similar degree of philosophical complacency and thoughtlessness can often be detected when research works and other materials are reproduced and made available by those transnational corporations associated with disruptive new media technologies, including social and mobile media, e-books, search engines, and the cloud. Writing on his Occupy 2012 blog, for example, Nicholas Mirzoeff begins a post on Michael Hardt and Antonio Negri's "Declaration" in "the manner of Derrida in *Limited Inc.*, ... with the inside matter." He does so to tease the neo-Marxist authors of *Multitude* and *Commonwealth* for having published their pamphlet on the global social movements of 2011 with Amazon using a "Copyright ... All rights reserved" license. "For a project about commoning, wouldn't a copyleft or Creative Commons license be more appropriate?" Mirzoeff asks. "OK, it's only 99 cents on Amazon but you have to have a Kindle-friendly device: why not just put out a free PDF?"[22]

No doubt for many there *is* something hypocritical about radical theorists advocating a politics of the Commons, commoning and communism, yet appearing to let little of this politics have an impact on the decisions they make (or that are made for them) regarding their own work, business, role, and practices as authors. And all the more so when a good number of them end up supporting "feral," profit-maximizing commercial publishers as a result. Amazon, for instance, is among those privately owned—some call them pirate capitalist[23]—companies that have aggressively avoided paying the standard rate of corporation tax in the United Kingdom (which was 26 percent in 2011, 24 percent in 2012, and stands at just 20 percent at the time of writing in 2015) along with Apple, Facebook, Google, Starbucks, and Uber. (Amazon only began paying corporation tax on its UK retail sales in May 2015—before that they were recorded in Luxembourg.) If there are calls for a postcrash economics, a radical rethinking of the field of economics that would challenge its own foundational assumptions in the light of the most recent crisis of capitalism, it is hard not to conclude we need a postcrash critical theory too.[24]

My concern, however, is not to develop a moralistic critique of such erstwhile radical thinkers for failing to make their knowledge and research available on a copyleft, Creative Commons, or open access basis. After all, none of these stances are necessarily anticommercial or anticapitalist either. What is so interesting about the question of the politics of sharing online when it is approached in relation to theory is its potential to raise the stakes even higher than Mirzoeff's commentary on "Declaration," which he hopes "isn't just a cheap shot." The kind of philosophical irresponsibility I am referring to extends even to those (still too rare) occasions when theorists *do* attempt to make their work openly available for others to copy, distribute, sell, or reuse. Such is the tendency of many theorists to rely on predefined—and sometimes only superficially understood—ideas of copyleft, Creative Commons, open access, and open source and of the differences between them when doing so, that they often get caught up in replicating uncritically, whether knowingly or not, many of the established concepts, values, methods, and practices to do with the individualized author, originality, fixity, the book, the human, and so on that these movements and initiatives themselves presuppose and take for granted.

To illustrate what I mean as far as the author, originality, fixity, and the human are concerned, let me take as an example a current of contemporary philosophy that is often regarded as offering a fundamental challenge to the central place of the human in Western thought.[25] In "The Importance of Bruno Latour for Philosophy," the object-oriented philosopher Graham Harman positions Immanuel Kant's critical shift of attention away from the world itself and onto "the conditions of possibility of human access to the world" as *the* major event of modern thought.[26] With this change of focus, Kant is seen as having "enslaved philosophy to a mighty central rift between human awareness and whatever may or may not lie outside it" (32). The result is a situation in which language is regarded as a problem of central importance to philosophy, but the relation between subway trains and steel is not. For Harman, writing in 2007, it is this state of affairs that explains why Latour, for all he regards himself as a philosopher and metaphysician, remains "almost invisible to academic philosophy" (31): because he "flatly rejects the single unique correlation between humans and world that abandons non-human objects to the calculating supervision of natural science" (32). Harman therefore takes it on himself to establish a philosophical reputation for Latour. He does so by positioning Latour at the "center of an unrecognized third strand in contemporary philosophy," *School X* (32), as Harman calls it, referring to the title of an "unpublished manuscript, in preparation" (48, n. 3) in which he also shoulders this task. In this third strand of philosophy (the other two strands are the analytic and continental schools), one of the basic assumptions of modern thought exits the stage:

> This assumption is that the relational gap between humans and world (whether we mourn it, revel in it, deconstruct it, or sublate it into some deeper absolute) is the sole gap with which philosophers have permission to be concerned. Latour outflanks this tiresome, oppressive, and often invisible dogma by reminding us that the relation between Immanuel Kant and the objects in the world is no different in kind from those between police and criminals, Lucky and Godot, reindeer and forests, acid and metal, or fire and cotton. Every actant has equal rights in a democratic ontology, and relations are a problem for all of them—not just for so-called rational beings. (43)

In short, Harman presents Latour as having given us "possibly the first object-oriented philosophy" on the grounds that there is "no privilege for a unique human subject" as far as the latter is concerned. "Instead, you and I are actants, Immanuel Kant is an actant, and dogs, strawberries, tsunamis, and telegrams are actants. With this single step," Harman writes, "a total democracy of objects replaces the long tyranny of human beings in philosophy" (36).

Interestingly, when Harman comes to publish *Prince of Networks: Bruno Latour and Metaphysics* two years later, he does so on an open access basis with re.press, using precisely the kind of Creative Commons license that would no doubt be considered by many to have been more suitable for Hardt and Negri's "Declaration."[27] Yet although it is available open access, this does not mean a network of people, objects, or actants can take Harman's text, rewrite and improve it, and in this way produce a work derived from it that can then be legally published. Since Harman has chosen to publish *Prince of Networks* using a Creative Commons attribution, noncommercial, no derivatives (BY-NC-ND) license, the most restrictive of CC's main licenses, any such act of rewriting would infringe his claim to copyright.[28] This applies to the right Harman wishes to retain to be identified as the author of *Prince of Networks* and to have it attributed to him precisely as a unique human subject. It also applies to Harman's right of integrity, which enables him as a singular human being to claim the original ideas it contains as his intellectual property and which grants him the privilege of refusing to allow the original, fixed, and final form of *Prince of Networks* to be modified or distorted by others, be they humans, objects, or nonhumans such as dogs, strawberries, tsunamis, and telegrams. Nor is this a situation that could have been avoided if Harman had chosen a less restrictive Creative Commons license for *Prince of Networks,* such as the Attribution 3.0 (CC-BY) license that "The Importance of Bruno Latour for Philosophy" was published under in the open access journal *Cultural Studies Review*.[29] For while a CC-BY license would mean *Prince of Networks* could be remixed, transformed, and built on by others just so as long as any changes made were indicated, Harman would still be retaining the right to be identified as the book's original author and to have it attributed to him precisely as *a* unique human subject. As a result, for all that his object-oriented philosophy is concerned to displace the human and the subject from the center of Western thought, and for all he has published, both *Prince of Networks* and "The Importance of Bruno Latour for Philosophy," on an open access basis using Creative Commons licenses—indeed precisely because he has done so—we can see that when it comes to his own work, business, role, and practice as a philosopher, Harman continues to adhere to a post-Enlightenment conception of the individual human subject as a legitimate holder of rights and property as much as any of those modern philosophers he chastises for having continued in the tradition of Kant.[30]

Granted, given the dearth of legal, economic, and political antihumanist alternatives to publishing either on a "Copyright … All rights reserved" or open access and Creative Commons basis that are professionally recognized, there is no quick or easy way of responding to this raising of the stakes for critical theory and philosophy. One thing is for sure: such contradictions and paradoxes are far from confined to Harman—or Latour for that matter, who continues to act as if he is a modern in this respect, even as he insists that *we have never been modern.*[31] In fact (and as we shall see), difficulties of this kind affect the majority of those who are currently endeavoring to overcome the tyranny of the human by highlighting the importance of the nonhuman, the posthuman, the postanthropocentric, and the multiscalar logic of the Anthropocene instead. Thanks to the way in which they too have responded to the question of the politics of copying, distributing, selling, and reusing theory and philosophy, such "posttheory theories" continue to be intimately caught up with the human in the very enactment of their attempt to think through and beyond it.[32]

Pirate Philosophy

Be that as it may, the high stakes that are raised for theory and philosophy by this question remain, for the point I am making here is also more than a cheap shot. So let me start to draw this opening account of the politics of sharing and the Commons to a close with a further question, one that is also a plea: How can we operate differently with regard to our own work, business, roles, and practices to the point where we actually begin to confront, think through, and take on (rather than take for granted, forget, repress, ignore, or otherwise marginalize) some of the implications of the challenge that is offered by theory to fundamental humanities concepts such as the human, the subject, the author, the book, copyright, and intellectual property, for the ways in which we create, perform, and circulate knowledge and research? Might acting as something like *pirate philosophers* be one way for us to do so?

In using the word *pirate* here, I have in mind not so much the historical romantic outsider of fiction and film (Captain Blood, Captain Jack Sparrow) or the kind of radical libertarian represented for some by the online drug czar Dread Pirate Roberts, founder of the underground drug site Silk Road;[33] or even the deviant thief or subversive radical associated with the anticopyright advocates of the pro-piracy movement (the Pirate Party and so on). When the word *pirate* first begin to appear in the texts of the ancient Greeks, it was "closely related to the noun *peira*, "trial" or "attempt," and so to the verb *peiraō*: the "pirate" would then be the one who "tests," "puts to proof," "contends with," and "makes an attempt."[34] It is these etymological origins of the modern term I am thinking of primarily, although I am aware the words for teasing (*pi̱ragma*) and giving trouble (*pira̱zo*) in modern Greek are both derived from that for pirate (*pirati̱s*) too.

It is this question, of how we can work toward the development of what might, given the fundamental importance of the human, the subject, the author and copyright to the humanities, also be thought in terms of *posthumanities*, which the chapters in this book in their different ways all address. They do so not so much with a view to resolving or even avoiding its many ambiguities and anxieties, problems and paradoxes, as finding ways of enduring and living with them by putting them to the test and teasing out some of their productive elements and dynamic potentialities. (And this includes those paradoxes and potentialities that are associated with my own apparent inability to simply transcend the "I" here and consequent production of a book with only my name on it about the problems involved in authors' producing books with only their names on them, even though I am aware *Pirate Philosophy* is written by what, for shorthand, can be referred to as the other[s] in me.) Of course, in a book of this size, it is impossible to engage with all of these inherited humanities concepts, values, forms, and practices in the same amount of depth and detail. Although they are interrelated—each of the chapters in *Pirate Philosophy* touches on most of these issues to some degree—decisions of emphasis have had to be made nonetheless. Due to their importance, both to our current ways of being theorists but also to any new ways of being we might endeavor to experiment with and actualize, I have focused on the Commons and community in chapter 1, the (future of the) humanities in chapter 2, the human subject in chapter 3, the posthuman and posthumanities in chapter 4, copyright and piracy in chapter 5, and the book in chapter 6.

That said, *Pirate Philosophy* does not constitute an attempt on my part to invent an overarching theory or seamless philosophical system: one that is then set out consistently to run right through this book in a logical, linear fashion from chapter to chapter, binding it tightly together into a unified, homogeneous whole. As we have seen, the politics of sharing and of knowledge (post-)production does not have fixed or predetermined meanings. Rather, it is continually being generated within an extended meshwork of dynamic flows and interweaving relations concerning the human, the subject, the author, the law, the market economy, and so forth that constitutes our contemporary material, social, and institutional environment. It is a politics that can thus differ significantly from place to place and from time to time. Accordingly—and as my reference to both pirate philosophy and the posthumanities suggests—rather than featuring one big idea, this book is multithemed and polycentered. When it comes to considering how we can operate otherwise with regard to our own work, business, roles, and practices, each chapter in *Pirate Philosophy* seeks to (temporarily) unbind a particular spatial and temporal knot in this extended meshwork. Chapter 2 does so through a close reading of Lev Manovich's quantitative cultural analysis in relation to the "computational turn" to data-led methods in the humanities (sometimes called *big humanities*); chapter 3 by focusing on Bernard Stiegler's philosophy of technology and time in the context of the cultural and program industries of the twenty-first century; chapter 4 by engaging with Rosi Braidotti's theory of the posthuman in respect of debates around open source, open science, and open access; chapter 5 by addressing certain phenomena associated with Internet piracy such as that of Napster, the Pirate Bay, and Aaaaarg; and chapter 6 by means of a speculative account of the future of the book in an era in which texts are generally connected to a network of other information, data, and mobile media environments. (Consequently, with a little adjustment, *Pirate Philosophy* could quite easily have had a title centering it on the digital humanities, capitalist subjectivation, the posthuman, posthumanities, or even the unbound book.)

Its heterogeneous and polycentric nature also helps to explain why, although my argument is constantly on the move within the chapters that constitute *Pirate Philosophy*, its overall development between chapters is at times less straightforward, more staggered, and indirect. Indeed, when read linearly, the argument played out over the course of *Pirate Philosophy* often tacks back to refocus on concepts, passages, ideas, and issues covered in previous chapters, the analysis of which may nevertheless be subject to a certain degree of change and transformation, as the various treatments of both the digital humanities and open access demonstrate. That some of the implications of the shift in political mood post-2008 are used as a starting point for the different treatments of the politics of copying, selling, and reusing theory that take place in chapters 1, 3, and 5, is a further example of this recursive, reiterative aspect of the book's flow of ideas. The chapters that make up *Pirate Philosophy* thus speak to and interact with one another in dynamic, intertwined, and at times perhaps surprising ways.

Pirate Philosophy is heterogeneous in its methodology too. Its theoretical models alone include the new materialism of Rosi Braidotti, the "new critique of political economy" of Bernard Stiegler, the control society thesis of Gilles Deleuze, the deconstruction of Jacques Derrida, the "poststructuralism" of Jean-François Lyotard, the anthropology of Tim Ingold, and the Cultural Analytics of Lev Manovich. For reasons that will become clear (see chapter 5), I do not intend to make too much of the pirate trope in connection to either historical or contemporary pirates, at least as they are conventionally conceived and understood. But if I did want to draw an analogy with the desire of both to challenge hierarchies of authority, then this book could even be said to be leaderless to the extent it is not overseen by the philosophy of a master thinker (be it Marx, Deleuze, Latour, Haraway, Laruelle, or whoever else): someone whose ideas it uses to secure its authority and align it with fellow disciples who display their allegiance through the sharing of the same references and metalanguage. Instead—and in keeping with the point I made earlier about thinking the nature of community, of being together and having something in common, differently—*Pirate Philosophy* holds the community of disparate and at times contradictory and even incommensurable theories it contends with and puts to the proof together in a productive, if at times uneasy, tension—so much so that chapter 4 practices simultaneously two traditions of thought often positioned as being antagonistic to one another: that based on a processual and relational ontology, developed through the writings of Baruch Spinoza, Henri Bergson, Gilles Deleuze, and Félix Guattari, and that associated with the philosophy of the other of Emmanuel Lévinas, Jacques Derrida, Ernesto Laclau, and Chantal Mouffe, and their emphasis on the responsible ethical and political decision.

All this antagonism, indirection, reiteration, and polycenteredness is deliberate. It relates to the importance that is placed on zigzagging, nonlinear, rhizomatic thinking by Braidotti (chapter 4). But it is also designed to complicate and make that little bit more difficult—or at any rate to not simply go along with[35]—any attempt to stabilize and solidify the theory that is being performed here (albeit in singular and thus somewhat different ways in each chapter) into a would-be fashionable intellectual brand in its own right: a big, new, "masculine," explanatory, and empowering theory of pirate philosophy or the posthumanities that can be set up in a relation of competition and rivalry to other big, explanatory, and empowering theories and philosophical systems, such as those associated with the posthuman, new materialism, media archaeology, object-oriented philosophy, speculative realism, and nonphilosophy (a theory that could be too easily sold, blogged, and tweeted about as *my* original work, intellectual property, or trademark in order to reinforce *my own* expertise and position in the academic marketplace, and thereby gain advantage in the struggle for attention, recognition, fame, authority, and disciplinary power).[36] Instead, the narrative structure and methodology of this book are intended to emphasize that any enactment of something like a pirate philosophy is possible only by virtue of a certain process of careful reading, interpreting, and thinking through—and thus getting close experience of, trying, testing, teasing, and troubling—the theories and philosophies of others. To put this in the language of a possible (and categorically plural) posthumanities, and as is demonstrated most clearly perhaps by the three longer chapters on the human, the humanities, and the posthuman that make up the first part of the book, *Pirate Philosophy* constitutes less something that belongs to me to the exclusion of all others and more of a critical, creative, and collaborative mixing and mutation—not least of the work of Manovich, Stiegler, Braidotti, and others.

The Art of Critique

The emphasis on the *critical* in the previous sentence is important and worth taking a moment to say more about in the light of recent tendencies in theory and philosophy. Another way of understanding the enactment of a posthumanities in this book is as a diffraction of the process of reading that Karen Barad, herself diffracting the insights of Donna Haraway, adopts as a critical tool. Barad does so in order to differentiate her own posthumanist performative approach from that of representationalism:

> Moving away from the representationalist trap of geo-metrical optics, I shift the focus to physical optics, to questions of diffraction rather than reflection. Diffractively reading the insights of feminist and queer theory and science studies approaches through one another entails thinking the "social" and the "scientific" together in an illuminating way. What often appears as separate entities (and separate sets of concerns) with sharp edges does not actually entail a relation of absolute exteriority at all. Like the diffraction patterns illuminating the indefinite nature of boundaries—displaying shadows in "light" regions and bright spots in "dark" regions—the relation of the social and the scientific is a relation of "exteriority within." This is not a static relationality but a doing—the enactment of boundaries—that always entails constitutive exclusions and therefore requisite questions of accountability.[37]

Now some have seen in this idea of a diffractive method of reading that attends to the relational nature of difference in neither spatial nor linear terms a means to produce "an analysis of the embeddedness of critique in that which it criticizes."[38] Others of what might be considered a more theoretically neurotic persuasion, however, have taken it as a cue to shift away from engaging with theoretical arguments in terms of what the philosophical tradition of Immanuel Kant, Theodor Adorno, and Michel Foucault is taken to understand as critique. Rather than finding faults, contradictions, and breaking points in the theories of others and arriving at negative judgments, they have gestured toward more constructive and assenting methods of working, seeing them as a means of generating affirmative practices capable of making a positive difference. Yet critique in the Kantian and Foucauldian sense cannot be set up against productive, affirmative approaches in a relation of contrast and opposition for the simple reason that such critique itself involves positive, creative production. As Foucault makes clear in his lecture addressing the question "What Is Critique?" critique is not reducible to arriving at negative judgments or standing against something and making opposing, corrective arguments. Critique is an art, a practice, a doing. It is "the art of not being governed like that and at that cost," as the first definition offered in his lecture has it—what a few pages later he refers to as the "art of voluntary insubordination." [39] Moreover, as Judith Butler acknowledges in her own critical reading of Foucault's text, critique is "a practice that not only suspends judgment ... but offers a new practice based on that very suspension."[40]

The constructive, creative, transformative aspect of critique deserves long and careful analysis. Suffice it to say for now that it is already apparent in the word's etymology, which reveals *critique* to be derived from the "Greek *kritike tekhne* 'the critical art.'"[41] (It can also be recognized in the way in which, for both Foucault and Butler, critique is used to identify the contingent conditions by which a field is constituted and transformed.) The reason for drawing attention to this constructive aspect of critique is that it has significant implications for the process of transforming the humanities into the kind of posthumanities performed in this book. Certainly there is an emphasis in *Pirate Philosophy* on being affirmative and creative. It is important, however, that such affirmation occurs in a nonoppositional relation to that which is also placed on reading the work of others critically in terms of the legacy of continental philosophy—and this includes analyses of the "exteriority within" and the embeddedness of critique in that which it criticizes. This emphasis on reading critically is vital, not least because it can help us avoid falling into the kind of traps identified in chapter 4. As is made clear there, to act other than critically, to not practice the art of critique and show "the complex interplay between what replicates the same process and what transforms it," risks the unwitting enactment of more or less the same old dismissive, negative, and dialectical structures of analysis that the theoretically neurotic shift away from critique is designed to elude.[42] The result can be an approach to critique that is neither particularly constructive, nor affirmative, nor critical. But reading critically is also crucial because as Foucault and Butler both emphasize in their different ways, critique entails "self-transformation" of the subject "in relation to a rule of conduct." It is thus "to risk one's very formation as a subject."[43] In other words, if we lack the courage to practice the art of critique, there is a danger of restricting ourselves primarily to the replication of what we already know and are and do.

A short section has been added at the end of each chapter in *Pirate Philosophy*, linking it to the next, to help orientate the reader. A certain amount of cross-referencing has been provided throughout the book for the same reason. In keeping with the emphasis in this chapter on heterogeneity and nonlinearity, however, the chapters that make up *Pirate Philosophy* can be read as singular interventions or cuts into our ways of being, thinking, and doing as theorists and philosophers in their own right, and in more or less in any order. For example, those who have time to read just one could pick chapter 5, "Pirate Radical Philosophy." Although it is not the final chapter, it is perhaps the nearest *Pirate Philosophy* gets to offering a summing up or conclusion. Meanwhile, those who—for all the points I have just made about diffractive methods of reading that attend to the relational nature of difference in neither spatial nor linear terms; about the *exteriority within* and *embeddedness of critique in that which it criticizes*; about mixing and mutating, and putting things to the *test* in order to *tease out* some of their productive elements and dynamic potentialities—are still keen to know more about how my argument in *Pirate Philosophy* relates to the construct known as *Gary Hall-the-theorist* I am enacting here, and to my own publication, with a university press, of a print-on-paper codex book complete with substantial references, quotations, and endnotes, might want to turn straight to the last chapter, "The Unbound Book."

Chapter 2 begins where the argument of this opening chapter leaves off: with the question of how we can operate differently with regard to our own work, business, roles, and practices. What I mean by this is how can we produce not just new ways of thinking but new ways of actually being theorists—ways that, among other things, do not necessarily have their basis in an extended authorship that is devoted to the building of arguments that are "comprehensive, monumental, definitive," and submitted to legacy journals and university presses (although they can do). I am referring to ways that as well as being "available to challenge, interruption and interpolation," are also hospitable to the experimental, processual, "provisional, ephemeral ... communal," to borrow from Stanley Fish's characterization of blogs and the digital humanities.[44] These are ways of being and doing as theorists that are even open to the idea of granting those involved—as some have claimed happens in the self-organized free universities that have proliferated in recent years—permission

> to start in the middle without having to rehearse the *telos* of an argument; to start from "right here and right now" and embed issues in a variety of contexts, expanding their urgency; to bring to these arguments a host of validations, interventions, asides, and exemplifications that are not recognisable as directly related or as sustaining provable knowledge. And, perhaps most importantly, "the curatorial," not as a profession but as an organizing and assembling impulse, opens up a set of possibilities, mediations perhaps, to *formulate subjects* that may not be part of an agreed-upon canon of "subjects" worthy of investigation.[45]

To this end, chapter 2 considers the extent to which the digital humanities do indeed offer us one productive way to think about such new ways of being.

2 THE HUMANITIES

There Are No Digital Humanities

What forms will critical theory take in the twenty-first century? Will the growing use of digital tools and data-led methods adopted from computer science and related fields to help analyze the vast, networked nature of knowledge and information in postindustrial advanced capitalist society produce a major change in theory and, indeed, the humanities? Certainly some of those associated with the digital humanities have suggested that we have already embarked on a post-theoretical era, exemplified by a shift away from a concern with ideology and critique and toward more quantitative and empirical modes of analysis. In this case, should critical theorists be looking to develop new forms of theory, characterized by an ability to combine the methodological and the theoretical, the quantitative and the qualitative, the digital and the traditional humanities? Are such new forms of theory and the humanities even possible? Or should we be looking to radically rethink what theory and the humanities are and what critical theorists do in the twenty-first century?

Part 1: On the Limits of Openness: The Digital Humanities and the Computational Turn

One of the interesting things about computer science is that, as Mark Poster emphasized some time ago, it was the first case where "a scientific field was established that focuses on a machine" and not on an aspect of nature or culture, as is the case with the physical, life, and social sciences. More interesting still is the way Poster was able to demonstrate that the relation to this machine in computer science is actually one of misrecognition, with the computer occupying "the position of the imaginary" and being "inscribed with transcendent status." His argument was that since "Computer Science found its first identity through its relation to the computer, that identity remains part of the disciplinary protocol of the field, even if the actual object, the computer, changes significantly, even unrecognizably, in the course of the years."[1] It is a misidentification on the part of computer science, however, that also has significant implications for any response we might make to the so-called computational turn in the humanities.[2]

Computational turn has been adopted to refer to the process whereby techniques and methods drawn from computer science and related fields—including interactive information visualization; science visualization; image processing; geospatial representation; statistical data analysis; network analysis; and the mining, aggregation, management, and manipulation of data—are used to create new ways of approaching and understanding texts in the humanities. Indeed, thanks to increases in computer processing power and its affordability in recent years, along with the enormous amount of cultural material now available in digital form, number-crunching software is being applied to millions of humanities texts in this way.

It is not my intention to equate this computational turn with the digital humanities.[3] Although the latter is sometimes known as humanities computing or as a transition between the traditional humanities and humanities computing, what has come to be called the digital humanities and this computational turn in the humanities should not be perceived as being equivalent.[4] Instead, I want to emphasize the importance of maintaining a distinction between them, especially if we are to develop a rigorous understanding of what the humanities can become in an era of networked digital information machines. So far (and as we shall see in part 2 of this chapter), the traffic in this computational turn has been too much one way. As the term implies, it has been concerned primarily with exploring what direct, practical uses computer science can be put to in the humanities in terms of performing operations on sets, flows, and networks of data so large that, in the words of the National Endowment for the Humanities' Digging into Data Challenge, "they can be processed only using computing resources and computational methods."[5] Witness Dan Cohen and Fred Gibbs's text mining of "the 1,681,161 books that were published in English in the UK in the long nineteenth-century," and Lev Manovich and the Software Studies Initiative's use of "digital image analysis and new visualization techniques" to study "20,000 pages of *Science* and *Popular Science* magazines ... , 780 paintings by Van Gogh, 4,535 covers of *Time* magazine (1923–2009)."[6] Just as interesting as what computer science has to offer the humanities, however, is the question of what the humanities, in both their digital and traditional guises (assuming the two can be distinguished in this manner, which is by no means certain, as we shall see), have to offer computer science. Beyond that, what can the humanities themselves bring to the understanding of computing and the shaping of the digital? Do the humanities really need to draw quite so heavily on computer science to develop a sense of what they can be in the age of new media, big data, and big analytics? Together with a computational turn in the humanities, might we not also benefit from more of a *humanities*—or, as I point to both in my conclusion to this chapter and in other chapters in this book, perhaps even post*humanities—turn* in our understanding of the computational and the digital?

Poster's argument about the relation to the machine in computer science being one of misrecognition takes on added importance in the light of such questions. It suggests that as a field, computer science is not necessarily the best equipped to understand itself and its own founding object, let alone help those in the humanities with their relation to computing and the digital.[7] In fact, counterintuitive as it may seem, if what we are looking for is an appreciation of what the humanities can become in an era of networked digital information machines and data-driven scholarship, we may be better advised to seek assistance elsewhere than from computer science and engineering, science and technology, or even science in general. One almost hesitates to suggest this in the current political climate, when so much government, research council, and private funding in the United States, UK, and elsewhere is focused on the STEM subjects (science, technology, engineering, and mathematics)—although it may be important to do so for just this reason. But perhaps we should turn to the theorists and philosophers of the humanities right from the start.

Three decades ago, the philosopher Jean-François Lyotard showed how science, lacking the resources to legitimate itself, had, since its beginnings with Plato, relied for this purpose on precisely the kind of knowledge it did not even consider to be knowledge: nonscientific narrative knowledge. Specifically, science justified itself by producing a discourse called philosophy. It was philosophy's role to generate a discourse of legitimation for science. Lyotard proceeded to define as *modern* any science that endeavored to prove itself credible in this way by means of a metadiscourse that explicitly appealed to a grand narrative of some sort: the life of the spirit, the Enlightenment, progress, modernity, the emancipation of humanity, the realization of the Idea. What makes Lyotard's *Report on Knowledge*, as it is subtitled, so significant with respect to the emergence of the digital humanities and the computational turn is that his ambition was not to position philosophy as being able to tell us as much, if not more, about science than science itself. It was rather to emphasize that in a process of transformation that had been taking place since at least the end of the 1950s, such long-standing metanarratives of legitimation had themselves become obsolete. So what happens to science when the philosophical metanarratives that legitimate it are no longer credible? Lyotard's answer, at least in part, was that science (or a certain stabilized, ideologically "accepted" version of it) was increasing its connection to society, especially the instrumentality and functionality of society (as opposed to, say, a notion of public service or common good).[8] Science was doing so by helping to legitimate and "augment" the power of states, companies, and multinational corporations by optimizing the "global relationship between input and output," between what is put into the social system and what is got out of it, in order to get more from less (46, 11).

It is at this point that we return directly to the subject of computing. For Lyotard, writing in 1979, technological transformations in research and the transmission of acquired learning in the most highly developed societies, including the widespread use of computers and databases and the "miniaturization and commercialization of machines," were already in the process of exteriorizing knowledge in relation to the "knower" (4). He demonstrates how this general transformation and exteriorization is leading to a major alteration in the status and nature of knowledge: away from a concern with "the true, the just, or the beautiful, etc." (44), with ideals (48), with knowledge as an end in itself, and precisely toward a concern with improving the social system's performance, its efficiency (xxiv)—so much so that for Lyotard,

> The nature of knowledge cannot survive unchanged within this context of general transformation. It can fit into the new channels, and become operational, only if learning is translated into quantities of information. We can predict that anything in the constituted body of knowledge that is not translatable in this way will be abandoned and that the direction of new research will be dictated by the possibility of its eventual results being translatable into computer language. The "producers" and users of knowledge must now, and will have to, possess the means of translating into these languages whatever they want to invent or learn. Research on translating machines is already well advanced. Along with the hegemony of computers comes a certain logic, and therefore a certain set of prescriptions determining which statements are accepted as "knowledge" statements. (4)

Some thirty years later, we do indeed find numerous discourses in the sciences taken up with exteriorizing knowledge and information in order to achieve "the best possible performance" by eliminating delays and inefficiencies and by solving technical problems (77). Thus we have John Houghton's study showing that the open access (OA) academic publishing model championed most vociferously in the sciences, whereby peer-reviewed scholarly research and publications are made available for free online to all those who are able to access the Internet, without the need to pay subscriptions to view it, is actually the most cost-effective mechanism for academic publishing.[9] Others have detailed the increases that open access publishing and the software related to it make possible in the amount of research material that can be published, searched, and stored; the number of people who can access it; the impact of that material; the range of its distribution; and the speed and ease of reporting and information retrieval—facilitating what Peter Suber, one of the leaders of the open access movement, has referred to as "better metrics."[10] Even the data sets created in the course of scientific research are being made freely and openly available on the Internet for others to use, analyze, and build on. Known as open data, this initiative is motivated by more than an awareness that data are the main research outputs in many fields. In the words of Alma Swan, another of the leading advocates for open access, publishing data sets online on an open basis bestows them with a "vastly increased utility": digital data sets are "easily passed around"; they are "more easily reused," reanalyzed, and checked for accuracy and validity; and they contain more "opportunities for educational and commercial exploitation."[11]

In a further move in this direction, some academic publishers are viewing the linking of their journals to the underlying data as another of their "value-added" services to set alongside automatic alerting and sophisticated citation, indexing, searching, and linking facilities (and to help ward off the threat of disintermediation posed by developments in digital technology that make it possible for academics to take over the means of dissemination and publish their work for and by themselves, cheaply and easily). All Public Library of Science (PLoS) open access journals, for example, now provide a broad range of article-level metrics and indicators relating to usage data on an open basis. No longer withheld as trade secrets, these metrics indicate which articles are attracting the most views, citations from the academic literature, social bookmarks, coverage in the media, comments, responses, "star" ratings, blog coverage, and so on. PLoS positions this program as enabling science scholars to assess "research articles on their own merits rather than on the basis of the journal (and its impact factor) where the work happens to be published," and they encourage readers to carry out their own analyses of this open data.[12] Yet it is difficult not to see article-level metrics as also being part of the wider process of transforming knowledge and learning into "quantities of information" that are produced more to be exchanged, marketed, and sold—for example, by individual academics to their departments, institutions, funders, and governments in the form of indicators of "quality" and "impact"—than for their "'use value.'"[13] Indeed, according to Roger Burrows, it would be "quite easy to generate a list of over 100 different nested measures to which each individual academic in the UK is now (potentially) subject."[14]

Certainly the requirement to have visibility, to show up in the metrics, to be measurable, encourages researchers to publish as much and as frequently as they can. The peer-reviewed academic journal article is consequently positioned by some as having now assumed "a single central value, not that of bringing something new to the field but that of assessing the person's research, with a view to hiring, promotion, funding, and, more and more, avoiding termination."[15] In such circumstances "it is not hard to visualize learning circulating along the same lines as money, instead of for its 'educational' value or political (administrative, diplomatic, military) importance."[16] Just as money has become a source of virtual value and speculation in the era of American-led neoliberal global finance capital, so too have education, research, and publication.

Such discourses around openness, efficiency, and utility are not confined to the sciences or even to the university. There are also wider political initiatives, dubbed "Open Government" or "Government 2.0," with first the Labour and then the Conservative/Liberal Democrat coalition administrations in the UK making a great display of freeing government information. The former implemented the Freedom of Information (FOI) Act in 2000 (subsequently regarded as a mistake and as a stick with which to beat political leaders by Tony Blair, prime minister at the time). In January 2010, Labour launched a website, www.data.gov.uk, expressly dedicated to the release of governmental data sets, a website the Conservative/ Liberal Democrat coalition continued to use extensively. Like the current Conservative government, the latter perceived transparency and open data very much as a means of driving economic growth—so much so that in 2012, the government established the Open Data Institute, codirected by Sir Tim Berners-Lee, expressly designed to build on the demand for open data.

Nor is this a phenomenon restricted to the UK; if anything, the situation is even more intense in the United States. Throughout his first presidential election campaign, Barack Obama repeatedly promised to make government more open. He followed this up by issuing a memorandum on transparency his first day as president in January 2009, in which he pledged to make openness one of "'the touchstones of this presidency'":[17] "My Administration is committed to creating an unprecedented level of openness in Government. We will work together to ensure the public trust and establish a system of transparency, public participation, and collaboration. Openness will strengthen our democracy and promote efficiency and effectiveness in Government."[18] How much he honored this commitment is highly questionable, especially in light of the US government's subsequent reaction to WikiLeaks: condemning the whistle-blowing website as "reckless and dangerous" in November 2010 after it opened up access to hundreds of thousands of State Department documents;[19] putting pressure on Amazon, PayPal, and others to stop supporting WikiLeaks—Amazon's response being to remove it from their servers on the first of December that year—and imprisoning Corporal Chelsea Manning for releasing the original classified material. Nevertheless, whereas in the UK a serving secretary of state (Mo Mowlam) could conceal a malignant tumor from both the public and the prime minister, such is the emphasis on freedom of information in the United States that knowledge of President Obama's resting heart rate (56 beats a minute), blood pressure (105/62), and cholesterol level (54.mmol/liter) is publicly available.[20]

From a liberal democratic perspective, freeing publicly funded and acquired information and data—whether gathered directly in the process of census collection, or indirectly as part of other activities (crime, health care, transport, schools, and accident statistics)—is indeed seen as helping society perform more efficiently. Openness is said to play a key role in increasing citizen trust, participation, and involvement in democracy, and government as access to information, such as that needed to inform and intervene in public policy, is no longer restricted to the state or to corporations, institutions, agencies, and individuals with sufficient money and power to acquire and possibly monopolize it for themselves. Such beliefs find support in the idea that making information and data freely and openly available goes along with Article 19 of the Universal Declaration of Human Rights, which states that everyone has the right "to seek, receive and impart information and ideas through any media and regardless of frontiers."[21] In 2010 US Secretary of State Hillary Clinton put forward a similar vision when she said that the United States stands "for a single internet where all of humanity has equal access to knowledge and ideas" against the authoritarian censorship and suppression of free speech and search facilities online and persecution of Internet users in countries such as China and Iran:

Even in authoritarian countries, information networks are helping people discover new facts and making governments more accountable.

... And technologies with the potential to open up access to government and promote transparency can also be hijacked by governments to crush dissent and deny human rights.

Some countries have ... violated the privacy of citizens who engage in non-violent political speech. These actions contravene the Universal Declaration on Human Rights.[22]

Of course, there is a significant difference between open access and open government: governments are perfectly able to advocate on behalf of openness while not being particularly open themselves. Nor does merely making information and data available to the public online on a transparent basis in itself necessarily change anything. In fact, such processes are often adopted precisely as a means of avoiding change. Aaron Swartz provides the example of Watergate, after which "people were upset about politicians receiving millions of dollars from large corporations. But, on the other hand, corporations seem to like paying off politicians. So instead of banning the practice, Congress simply required that politicians keep track of everyone who gives them money and file a report on it for public inspection."[23] Yet besides appearing rather ironic, particularly after the Snowden revelations of June 2013 concerning the National Security Agency, PRISM, TEMPORA, and XKeyscore surveillance programs,[24] and the more recent scandal over her own use of a private email account for official public business, a known tactic for eluding public records requests, is Clinton not also guilty in this speech of overlooking (or conveniently forgetting and even denying) the way liberal ideas of freedom and openness (and of the human) have long been used in the service of imperialism, colonialism, and neoliberal globalization? Does freedom for neoliberal globalization not primarily mean economic freedom, freedom of the market, freedom of consumers to choose what to consume—not only in terms of goods but also lifestyles, ways of being? Even if the data come from an era before the widespread use of networked computers, it is interesting that "fifteen years after the Freedom of Information Act [FOIA] law was passed" in the United States in 1966, "the General Accounting Office reported that 82 percent of requests [for information] came from business, nine percent from the press, and only 1 percent from individuals or public interest groups."[25] Certainly, as far as the United Kingdom of the twenty-first century is concerned, the "truth is that the FOI Act isn't used, for the most part, by 'the people,'" as Tony Blair acknowledges in his memoir. "It's used by journalists"—and, one might add, by privately owned companies.[26]

In view of this, it is no surprise that neoliberal conservatives also support making the data freely and openly available to businesses and the public on the grounds that doing so provides a means of achieving what Lyotard refers to as the "best possible input/output equation" (46). In this respect, it is of a piece with the emphasis placed by neoliberalism's audit culture on accountability, transparency, evaluation, measurement, and centralized data management. Such openness and communicative transparency are perceived as ensuring greater value for (taxpayers') money, helping drive up standards, eliminate corruption (e.g., in the UK over expense payments for second homes for members of Parliament—although most references to these were redacted out of FOIA requests and came to light only as a result of leaks), enabling costs to be distributed more effectively, and increasing not just choice, competiveness, and accountability, but enterprise, creativity, and innovation too, thus leading to economic and social growth. Well-financed companies—including private technology firms, policy labs, and research and development labs that, unlike the majority of the public, have the time, resources, and expertise to exploit (and enclose) these publicly available data sets—are able to use them to build new businesses by creating data analytics, expanding existing markets, and generating new markets and services.[27] The monetization of personal data is already reported to be a $156-billion-a-year industry in the United States.[28] The potential to generate billions for the economy in this manner was also one of the justifications behind the UK's care.data project to share the health records of National Health Service patients with private companies, medical prescription data "already being used by pharmaceutical firms to target investment in research."[29] The McKinsey Global Institute even goes so far as to insist that in the future, "analyzing large data sets—so-called big data" will be a "key basis of competition, underpinning new waves of productivity growth, innovation, and consumer surplus."[30]

Yet to have participated in this shift that the widespread use of computers and databases has helped to bring about, away from questions of truth, justice, and especially what Lyotard elsewhere places under the headings of "heterogeneity, dissensus, event … the unharmonizable," and toward a concern with improving the social system through an emphasis on performativity, measurement, and optimizing the relation between input and output, one does not need to have actively and consciously contributed to the movements for open access, open data, or open government.[31] As is well known now, if you are one of those who have helped Amazon at the time of this writing to sell an estimated $4.5 billion worth of Kindle-branded devices—the market for digital books now being larger than that for hardbacks on both Amazon's US and UK websites (the latter being itself not inconsiderable, given some are predicting Amazon will soon account for half of all book sales in the United States)[32]—you have signed a license agreement allowing the online book-retailer-turned-technology-company, *but not academic researchers or the public*, to collect, store, mine, analyze, and extract economic value from data concerning your personal reading habits for free. This includes what, where, when, and how much you read (or don't read), as well as any notes, highlights, and underlining you add to the text. Similarly, if you are among the 1.39 billion people worldwide who have joined Facebook's password-protected "walled garden" social network, you have voluntarily donated your time and labor to help its owners and their investors generate $48 billion of revenue from demographically targeted mobile advertising—and that's in the fourth quarter of 2014 alone.[33] (Facebook was valued at over $104 billion on the day of its initial public offering in May 2012.) Even if you have done neither, you have in all probability provided AOL, Google, Microsoft, Skype, Yahoo, or YouTube with a host of free information and data relating to you, your family, friends, colleagues, peers, and correspondents that they can monitor, monetize, and, as we know following Snowden, give—whether willingly or not, whether for our care, benefit, protection, and security or not—to governments and intelligence agencies such as the National Security Agency in the United States and the Government Communications Headquarters in the UK. As well as providing examples of horizontal Web 2.0 social networks or social media, then, Amazon, Facebook, and YouTube are very much top-down social hierarchies. They may challenge conventional notions of privacy (especially the case with regard to those that have developed in the West since the emergence of the book and the associated requirement for closed-off spaces in which to read and study that have played such an important role in the development of modern subjectivity and the public-private distinction). They may also be part of a widespread emphasis on the need for transparency. Yet they reserve a right to privacy and to a lack of transparency with regard to their own activities and business practices. (Google is likewise unhappy for anyone to photograph the exterior of their data farms, despite having devised both Google Street View and Google Earth.)[34] Consequently, while these companies and the governments that support them are popularly perceived as threatening the right to privacy,[35] what is less focused on is the way they are simultaneously engaged in the process of monopolizing the right to privacy for themselves—and perhaps in the future for anyone else who can afford to pay for the privilege.

Obviously, no matter how enjoyable such products and activities may be, no one has to buy a Kindle e-reader, join a social network, or display his or her personal metrics online, from sexual activity to food consumption, in an attempt to identify patterns in their life—what is referred to as life tracking, self-tracking or the "quantified self."[36] It should also be acknowledged that many people are quite happy to continue to be part of the networked communities reached by Amazon, Facebook, Google, and the others, even though they realize they are both contributing to large-scale surveillance and intelligence-gathering activities and being used as free labor. They just consider this part of the deal and as a reasonable trade-off for the often free-to-use services and experiences that these companies provide. (This attitude can be revised, however, if such services begin to appear too intrusive as a result of changes in privacy settings or the way updates and feeds are configured.) Nevertheless, refusing to be part of this move toward supplying ever more quantities of information, data, and work for free is not an option for most people. It is not something that can be opted out of or strategically abandoned and withdrawn from simply by declining to look for research using Google Scholar, removing the cookies from your computer, committing social networking suicide,[37] and reading print-on-paper artist books and zines bought with cash in a bricks-and-mortar artist-run bookshop instead.[38] In fact, it is not a question of actively doing something in this respect at all: of making yourself constantly available for data mining and surveillance, at your own time and cost, by contributing free labor to the likes of YouTube, Google+, Instagram, and Vine; of using only services that have a business model that does not rest on the exploitation of granular data for advertising and other purposes (e.g., because they charge a fee rather than being free to use, as in the case of the German anonymous e-mail provider Posteo); of challenging the data monopolies of Google, Facebook, and others and at the same time working to secure your economic future by becoming what is in effect a small-scale data-preneur, capturing, hoarding, managing, and selling your own personal data, now regarded very much as (your) property;[39] or of militantly refusing to interface with such digital control systems at all, perhaps even going offline completely and using a manual typewriter in the park.[40] As Gilles Deleuze and Félix Guattari pointed out a while ago, "surplus labor no longer requires labor ... one may furnish surplus-value without doing any work," or anything that even remotely resembles work for that matter, at least as it is most commonly understood:

> In these new conditions, it remains true that all labor involves surplus labor; but surplus labor no longer requires labor. Surplus labor, capitalist organization in its entirety, operates less and less by the striation of space-time corresponding to the physicosocial concept of work. Rather, it is as though human alienation through surplus labor were replaced by a generalized "machinic enslavement," such that one may furnish surplus-value without doing any work (children, the retired, the unemployed, television viewers, etc.). Not only does the user as such tend to become an employee, but capitalism operates less on a quantity of labor than by a complex qualitative process bringing into play modes of transportation, urban models, the media, the entertainment industries, ways of perceiving and feeling—every semiotic system. It is as though, at the outcome of the striation that capitalism was able to carry to an unequalled point of perfection, circulating capital necessarily recreated, reconstituted, a sort of smooth space in which the destiny of human beings is recast.[41]

In fact, the process of capturing data by means not just of the Internet but a myriad of communication satellites, eye-in-the-sky drones, cameras, sensors, and robotic devices is now so ubiquitous and pervasive that it is as good as impossible to avoid being unwittingly caught up in it, no matter how rich, knowledgeable, and technologically proficient you are. In 2011 the *Guardian* (in many respects leading the way in investigating such surveillance practices, even before its publication of the Snowden revelations) reported that there are approximately 1.85 million closed-circuit TV (CCTV) cameras in the UK—one for every thirty-two people. Yet no one really knows how many of those cameras are actually in operation in Britain today. (Indeed, the 1.85 million statistic is itself based merely on an extrapolation of a study of CCTV cameras in Cheshire. Moreover, while it is often said that the UK has the most surveillance cameras in the world, this claim cannot be verified without comparable research in other countries.)[42] And that is without even mentioning all the other means of gathering data that are reputed to be more intrusive still, such as mobile phone GPS location, the automatic vehicle number plate camera recognition system installed on the UK's major roads and in town centers to log 90 percent of vehicle journeys in real time, and the body-worn videos some police, university security staff, and even supermarket workers are wearing now. To avoid all unwanted data capture would require you, at the very least, to carry your phone in a radio frequency identification-blocking (RSID) metal-lined wallet; conceal your eyes, nose, and the bridge in between so you cannot be easily recognized by CCTV using facial recognition algorithms; and live for at least three years somewhere outside urban areas where there are fewer such cameras—a cave or tent in a remote location such as a forest would be preferable. (If someone were actively looking for you, you would have to remember not to leave any traces of your biological matter—chewing gum, cigarette ends, hairs, feces—that could be used to sequence your DNA and thus uncover your identity using biometrics.)

Yet going to these kinds of lengths is simply too exhausting and alienating for most people. And the more you try to hide or even just live in a manner that renders your data unavailable for algorithmic accumulation, specification, and classification—say, by refusing to carry a smart phone, encrypting your e-mails, using Tor to browse the Web anonymously, and flashing infrared beams into camera lenses—the more you draw the attention of the National Security Agency and other intelligence services to yourself anyway. Besides, such notions of militant refusal and active resistance (like their counterparts to do with ideas of freedom, privacy, civil rights, civil liberties, and even data-preneurship, and sousveillance)[43] too often have their basis in a conception of the rational, self-identical, and self-present individual humanist subject—precisely that which (as chapter 3 shows) is in the process of being reconfigured by these changes in media and technology. As a result, such gestures risk overlooking, or at best downplaying, and thus being unknowingly caught up by, the way computers, databases, archives, software, servers, blogs, image and video sharing, social networking, and the cloud are not just being used to change the status and nature of knowledge; they are involved in the constitution of a different form of human subject too.

To what extent do such developments cast the computational turn in the humanities in a rather different light to the celebratory data fetishism that has dominated much of this rapidly emerging field? Is the externalization of knowledge onto computers, databases, servers, and the cloud, and direct, practical use of techniques and methods drawn from computer science and various fields related to it, including management, business, and design, here too helping to produce a major alteration in the status and nature of knowledge—and indeed the humanities, humanists, and the human? One can think not just of the use of tools such as Anthologize, Delicious, Mendeley, Prezi, and Zotero to augment, structure, and disseminate scholarship and learning in the humanities; there is also the generation of dynamic maps of large humanities data sets, and employment of algorithmic techniques to search for and identify patterns in literary, cultural, and filmic corpora, as well as the way in which the interactive nature of much digital technology is enabling user data regarding people's creative activities with these media to be captured, mined, and analyzed by humanities scholars.

Certainly, in what seems to be almost the reverse of the situation Lyotard described, many of those in the humanities now appear to be looking increasingly to science (and technology and mathematics)—if not always computer science specifically—to provide their research with a degree of legitimacy. This includes some of the field's most radical thinkers. Witness Franco "Bifo" Berardi's appeal to "the history of modern chemistry on the one hand, and the most recent cognitive theories on the other" for confirmation of the compositionist philosophical hypothesis: "There is no object, no existent, and no person: only aggregates, temporary atomic compositions, figures that the human eye perceives as stable but that are indeed mutational, transient, frayed, and indefinable."[44] It is this hypothesis, derived from Democritus, that Bifo sees as underpinning the methods of both the schizoanalysis of Deleuze and Guattari and the Italian autonomist theory on which his own compositionist philosophy is based. ("Compositionism" is how Bifo prefers to speak of the stream of thought associated with Italian operaismo.) Can this turn toward the sciences be regarded as a response on the part of the humanities to the perceived lack of credibility, if not obsolescence, of their metanarratives of legitimation: the life of the spirit and the Enlightenment, but also Marxism, psychoanalysis, and so forth? Indeed, are the sciences today to be regarded as answering many humanities questions more convincingly than the humanities themselves?

While ideas of this kind are a little too neat and symmetrical to be entirely convincing, the scientific turn in the humanities has been attributed by some to a crisis of confidence brought about, if not by the lack of credibility of the humanities' metanarratives of legitimation exactly, then at least in part by the "imperious attitude" of the sciences. It is an attitude that has led the latter to colonize the humanists' space in the form of biomedicine, neuroscience, theories of cognition, and so on.[45] From this perspective, the turn toward computing appears as just the latest manifestation of, and response to, this crisis of confidence in the humanities. Can we go even further, however, and ask: Is it evidence that certain parts of the humanities are attempting to increase their connection to society and to the efficiency, instrumentality, and functionality of society especially?[46] What are we to make of the fact that such a turn toward computing is gaining momentum at a time when the UK government is emphasizing the importance of business, management, and STEM and withdrawing support and funding for the humanities? No doubt it would require a long, complex, multifaceted analysis that goes some way back in history to answer this question. Still, one of the reasons all this is happening now may be due to the fact that the humanities, like the sciences themselves, are under pressure from government, business, management, industry, and increasingly big media to prove they provide value for money in instrumental, functional, performative terms. Can the interest in computing therefore be seen as a strategic decision on the part of some of those in the humanities? After all, one can get funding from the likes of Google and Twitter.[47] In fact, Patricia Cohen, in her article "Digital Keys for Unlocking the Humanities' Riches," reports that in the summer of 2010, "Google awarded $1 million to professors doing digital humanities research"; while in April 2014, Manovich announced that the Software Studies Lab had been awarded one of Twitter's inaugural #DataGrants to explore questions such as, "Can visual characteristics of images shared on social media tell us about the 'moods', 'happiness' and 'social well-being' of cities?" and "Do cities that are more 'happy' have more selfies and do people smile more when taking selfies?"[48]

At the very least, a question can be raised concerning the extent to which the adoption of practical techniques and approaches from computer science is providing some areas of the humanities with a means of defending and refreshing themselves in an era of global economic crisis and severe cuts to higher education, through the transformation of their knowledge and learning into (ideologically acceptable) quantities of information—deliverables. But the computational turn can also be positioned as an event created to justify such a move on the part of certain elements within the humanities.[49] In this case, it might be advisable to use a term different from *digital humanities* if we do not wish to simply go along with the current movement away from what remains resistant to a general culture of measurement and calculation. The idea of both the computational turn and the digital humanities seems to imply that, thanks to the development of a new generation of powerful computers and digital tools, the humanities have somehow become, or are in the process of becoming, digital.[50] Yet one of the things I am attempting to show here by drawing on the thought of Lyotard, Poster, and others is that the digital is not something that can now be added to the humanities for the simple reason that the (supposedly pre-digital) humanities can be seen to have already had an understanding of, and engagement with, computing and the digital (since at least 1979 in Lyotard's case).

After decades during which the humanities have been heavily marked by a variety of critical theories (Marxist, psychoanalytic, postcolonialist, post-Marxist), it is particularly noticeable how many instances of the turn to data-driven scholarship lack an awareness of computing and the digital as much more than a set of tools, techniques, and resources, and thus manifest as naive and lacking in meaningful critique.[51] Witness the emphasis on making the data not only visible but visual, even aesthetic. As Clare Birchall comments, the visible here "must be, as the title of David McCandless's bestselling book [*Information Is Beautiful*] suggests, beautiful. It is not enough ... to make data available, one must draw it out of the shadows of the deep web's darkest archives and use it to produce attractive data visualisations."[52] Stefanie Posavec's Literary Organism, which visualizes the structure of part 1 of Jack Kerouac's *On the Road* as a tree, provides another oft-cited example of this aestheticization of data (even though it was actually designed by hand, rather than using digital tools, as part of an attempt on Posavec's part to do things as a designer that computers are unable to do).[53]

There is a long history of critical engagement within the humanities with ideas of the visual, the image, the spectacle, the spectator, and so on—not just in critical theory but in literary studies, cultural studies, women's studies, queer studies, media studies, and film and television studies. Such a history of critical engagement stretches back to Guy Debord's influential 1967 work, *The Society of the Spectacle*, and beyond. For instance, in his introduction to *Visual Display: Culture beyond Appearances*, Peter Wollen writes that an excess of visual display within culture has "the effect of concealing the truth of the society that produces it, providing the viewer with an unending stream of images that might best be understood, not simply as detached from a real world of things, as Debord implied, but as effacing any trace of the symbolic, condemning the viewer to a world in which we can see everything but understand nothing—allowing us viewer-victims, in Debord's phrase, only 'a random choice of ephemera.'"[54] It can come as something of a surprise, then, to discover that this humanities tradition, in which ideas of the visual are engaged critically, appears to have had comparatively little impact on much of the enthusiasm for data visualization that is so prominent an aspect of the turn toward data-intensive scholarship.

This (at times explicit) repudiation of criticality can be viewed as part of what makes certain aspects of the digital humanities so intriguing at the moment. Exponents of the computational turn are endeavoring to avoid conforming to accepted (and often moralistic) conceptions of politics that have been decided in advance, including those that see it only in terms of power, ideology, race, ethnicity, gender, class, sexuality, and so on. Refusing to "go through the motions of a critical avant-garde," these champions are responding to what is perceived as a fundamentally new cultural situation, and the challenge it represents to our traditional methods of studying culture, by avoiding such conventional gestures and experimenting with the development of fresh methods and approaches for the humanities instead.[55] "Forget meaning," Timothy Lenoir tells N. Katherine Hayles. "Follow the datastreams."[56]

There may well be a degree of "relief in having escaped the culture wars of the 1980s"—for those in the United States especially—as a result of this move "into the space of methodological work"[57] and what Tom Scheinfeldt dubs "the post-theoretical age."[58] The problem, however, is that without such reflexive critical thinking and theories, many of those whose work forms part of this computational turn find it difficult to articulate the point of their contributions, as Scheinfeldt readily acknowledges.[59] Interestingly, Scheinfeldt suggests that the problem of theory, or the lack of it, may actually be a matter of scale and timing:

> It expects something of the scale of humanities scholarship which I'm not sure is true anymore: that a single scholar—nay, every scholar—working alone will, over the course of his or her lifetime ... make a fundamental theoretical advance to the field.
>
> Increasingly, this expectation is something peculiar to the humanities. ... It required the work of a generation of mathematicians and observational astronomers, gainfully employed, to enable the eventual "discovery" of Neptune. ... Since the scientific revolution, most theoretical advances play out over generations, not single careers ... There is just too much lab work to be done and data to [be] analyzed for each person to be pointed at the end point.[60]

Notice how theory is again marginalized in favor of an emphasis on STEM, and the adoption of expectations and approaches associated with mathematicians and astronomers in particular.

None of this is to deny that we should experiment with the new tools, methods, and materials that digital media technologies create and make possible, including those drawn from computer science, in order to bring new forms of Foucauldian *dispositifs,* or what Bernard Stiegler refers to as *hypomnémata* or *mnemonics* into play.[61] (It is certainly not my intention here to take part in a bout of "algorithm bashing," no matter how popular a sport it may be these days.) Still, there is something intriguing about the way many defenders of the turn toward computational tools and methods in the humanities evoke a sense of time in relation to theory. Take the argument, apparent in the emphasis Scheinfeldt places on scale and timing, that critical and self-reflexive theoretical questions about the use of digital tools and data-led methods should be deferred for the time being, lest they have the effect of strangling at birth what could turn out to be a very different form of humanities research before it has had a chance to properly take shape. Viewed in isolation, it can be difficult, if not impossible, to decide whether this particular kind of limitless postponement is serving as an alibi for a naive and rather superficial form of scholarship[62] or whether it is indeed acting as a responsible political or ethical opening to the (heterogeneity and incalculability of the) future, including the future of the humanities. After all, the suggestion is that now is not the right time to be making any such decision or judgment, since we cannot yet know how humanists will eventually come to use these tools and data, and thus what data-driven scholarship may or may not turn out to be capable of critically, politically, and theoretically. This argument would be more convincing as a responsible political or ethical call to leave the question of the use of digital tools and data-led methods in the humanities open, however, if it were the only sense in which time was evoked in relation to theory in this context. Significantly, it is not. As we have seen, advocates for the computational turn do so in a number of other and often competing senses too:

1. That the time *of* theory is over, in the sense that a particular historical period or moment has now ended (for example, that of the culture wars of the 1980s)
2. That the time *for* theory is over, in the sense that it is now the time for methodology[63]
3. That the time to return to theory, or for theory to (re-)emerge in some new, unpredictable form that represents a fundamental breakthrough or advance, although possibly on its way, has not arrived yet and cannot necessarily be expected to do so for some time (given that "most theoretical advances play out over generations")[64]

All of this gives a very different inflection to the view of theoretical critique as being at best inappropriate and at worst harmful to data-driven scholarship. Even a brief glance at the history of theory's reception in the English-speaking world is sometimes enough to reveal that those who announce its time has not yet come or is already over, that theory is in decline or even dead, and that we now live in a post-theoretical world, are more often than not endeavoring to keep it at a temporal distance. Positioning their work as either pre- or post-theory in this way in effect grants them permission to continue with their preferred techniques and methods for studying media and culture relatively uncontested (rather than having to ask rigorous, critical, and self-reflexive questions about their own practices and justifications for them). Placed in this wider context, far from helping keep open the question concerning the use of digital tools and data-led methods in the humanities, the rejection of critical-theoretical ideas as untimely can be seen as both moralizing and conservative.

In saying this, I am reiterating an argument made by Wendy Brown in the sphere of political theory. Yet can a similar case not be made with regard to the computational turn in the humanities, to the effect that the "rebuff of critical theory as untimely provides the core matter of the affirmative case for it?"[65] Theory is vital from this point of view, not for conforming to accepted conceptions of political critique that see it primarily in terms of power, ideology, race, ethnicity, gender, class, and so on, but "to contest the very sense of time invoked to declare critique untimely."[66]

Part 2: The Cultural Analytics of Lev Manovich and the Software Studies Initiative

To think further and in more detail about the relation between data-driven scholarship and theory and critique, let us turn to what has frequently been positioned as one of the most interesting and influential examples of the computational turn: the cultural analytics of Lev Manovich and the Software Studies Initiative.[67] For Manovich, it is not simply a matter of the widespread use of computers and databases exteriorizing knowledge in relation to the knower; as he makes clear in a series of essays, interviews, and other postings on the subject, it is a case of there now being so much cultural production in the twenty-first century that it can no longer be known by the knower. In 2012, for instance, there were already "2.2 billion email users worldwide ... 634 million websites ... 2.7 billion likes on Facebook every day, 175 million tweets ... sent ... every day, 4 billion hours of video ... watched on YouTube monthly."[68] Manovich thus sees the sheer scale and dynamics of this new media landscape as presenting the accepted means of studying culture—the kind of theories, concepts, and methods appropriate to producing close readings of the content of a relatively small number of texts that were dominant for so much of the twentieth century, with a significant practical and conceptual challenge. In the past, "cultural theorists and historians could generate theories and histories based on small data sets"—the American literary canon of the 1960s, for example, or the films of Alfred Hitchcock. "But how can we track 'global digital cultures,'" Manovich asks, "with their billions of cultural objects, and hundreds of millions of contributors?"[69]

Manovich's solution to this data deluge is to turn to the very computers, databases, software, and vast amounts of born-digital networked cultural content that are creating the problem in the first place and use them to help develop new methods and approaches adequate to the task at hand. This is where what he variously calls Big Humanities, Quantitative Cultural Analysis, Cultural Datamining, or Cultural Analytics comes into play. While scientists, businesses, and government agencies are using data analytics to obtain not just figures from big data but also useful ideas for action (supermarkets, to provide another example to set alongside those I provided earlier, are selecting sites for the location of their new stores according to finely-grained analyses of the geographical and population data, with Clive Humby, inventor of Tesco's Clubcard loyalty card, quoted as early as 2006 as saying, "Data is the new oil"),[70] the "key idea of Cultural Analytics is the use of computers to automatically analyze cultural artifacts in visual media, extracting large numbers of features that characterize their structure and content."[71] And what is more, it does so not just with regard to the culture of the past but also with that of the present, including real-time data flows. To this end, Manovich calls for as much of culture to be made available in external, digital form as possible: "not only the exceptional but also the typical; not only the few 'cultural sentences spoken by a few 'great man' [*sic*] but the patterns in all cultural sentences spoken by everybody else."[72]

What makes Manovich and the Software Studies Initiative's Cultural Analytics research so interesting is the way it is clearly striving to open the humanities to some of the new disciplines, frameworks, and forms of knowledge digital media technologies may make possible:

> What will happen when humanists start using interactive visualizations as a standard tool in their work, the way many scientists do already? If slides made possible art history, and if a movie projector and video recorder enabled film studies, what new cultural disciplines may emerge out of the use of interactive visualization and data analysis of large cultural data sets?[73]

And, to be sure, Cultural Analytics is able to demonstrate some of the things software tools and quantitative analysis can do in this respect, particularly when it comes to identifying patterns, relationships, trends, tendencies, and structures in large sets of cultural data—or variations in, disruptions of, and exceptions to those patterns and trends. For example, they can be used to generate information such as the color "palettes of films as a whole," "the individual or group aesthetic impression of what typifies the 'essential' character of a film and ... which shots or scenes best correspond to that assessment," and can be visualized but not necessarily described in language.[74] Interactive visualizations of this kind may even have the potential to open up new directions in the analysis of film in terms of patterns, rhythms, and dynamic flows that change over time. Still, "visualization only shows patterns—it's up to the researcher to interpret them as meaningful."[75] Significantly, the role of actually interpreting such patterns as meaningful, let alone reflecting critically on the practice of doing so, is one Manovich frequently downplays, even marginalizes, in his accounts of Cultural Analytics. (How does the ascription of meaning to the underlying cultural patterns and relationships revealed by visualization avoid not just "the practice of apophenia: seeing patterns where none actually exist, simply because massive quantities of data can offer connections that radiate in all directions,"[76] but also becoming some kind of twenty-first-century new media formalism/structuralism?) He prefers to leave the task of critical interpretation to other researchers to undertake at some unspecified point in the future. "What we need is to have as many people as possible start using these tools—and then we will see what will emerge," he declares (evoking in doing so a sense of time and scale reminiscent of Scheinfeldt).[77] Consequently, what Cultural Analytics is not so clearly able to demonstrate—at least not yet anyway—is precisely the kind of rigorous critical interpretation and self-reflection that might open up new directions in the analysis of cinema, say, and turn all these large data sets and the information they produce into a new argument or hypothesis about culture.[78] It is often difficult to get a sense of what the resulting cultural criticism would look like from Manovich's descriptions of Cultural Analytics. What should the users of the open source tools Cultural Analytics wants to provide *do* with the results of their research?

To raise this issue is not to imply that some forms of quantitative cultural analysis or analytics cannot be used critically and self-reflexively to help explore and research the vast, networked nature of twenty-first-century postindustrial capitalist society[79] and even creatively analyze, subvert, resist, or reinflect culturally dominant discourses, including some of those associated with openness, efficiency, instrumentality, and transparency. Nor am I suggesting Manovich is going anywhere near as far in his data evangelism as Chris Anderson when the latter, with a breathless hyperbole typical of *Wired* magazine, argues that the life span of theory, including that associated with developing scientific models and hypotheses, is coming to an end as a result of being rapidly replaced by statistical algorithms:

> This is a world where massive amounts of data and applied mathematics replace every other tool that might be brought to bear. Out with every theory of human behavior, from linguistics to sociology. Forget taxonomy, ontology, and psychology. Who knows why people do what they do? The point is they do it, and we can track and measure it with unprecedented fidelity. With enough data, the numbers speak for themselves.[80]

A large part of the appeal of Manovich's particular enactment of the turn toward computing and data-driven scholarship (and this is partly why I have chosen to focus on his account of Cultural Analytics) lies with the way he continues to talk about asking "larger theoretical questions about cultures (as opposed to more narrow pragmatic questions" asked by professional fields associated with science, business, and government).[81] Manovich acknowledges that with Cultural Analytics, he wants to create tools "to enable new type [*sic*] of cultural criticism and analysis appropriate for the era of cultural globalization and user-generated media."[82] So, contra boyd and Crawford's characterization of many debates over big data, Manovich is not suggesting that "all other forms of analysis can be sidelined by production lines of numbers, privileged as having a direct line to raw knowledge," and that consequently we give up on critique and on asking theoretical questions.[83] Nevertheless, it is surprisingly hard to find actual instances where Manovich articulates in a rigorous fashion exactly how Cultural Analytics might be used to develop and perform such a new form of cultural criticism.

This difficulty on Manovich's part when it comes to delivering on the promise of his research to provide a new type of cultural criticism and analysis appropriate for the information society of the twenty-first century could lead his work to be regarded as something of a disappointment. It can certainly make engaging with his Cultural Analytics a frustrating experience. After all, as I say, Manovich expressly states that what he wants to do is both understand such digitizing processes and methods *and* subject them to critique. Indeed, the reason Cultural Analytics wants to understand them, according to Manovich, is precisely to be able to "critique them better."[84] Likewise, he insists, "we should be self-reflective. We need to think about the consequences of thinking of culture as data and of computers as the analytical tools: what is left outside, what types of analysis and questions get privileged, and so on. This self-reflection should be part of any Cultural Analytics study."[85] There is a significant difference between saying we need to be self-reflexive, however, and actually being self-reflexive.

It is worth noting that the origin of the word *data* is as the plural of the Latin word for *datum*, which means a proposition that is assumed, given, or taken for granted, often in order to construct a theoretical framework or draw conclusions. Moreover, in engineering, the datum point is the place from which measurements are taken. The datum point itself, however, is not checked or questioned; as the position from which measurements are made, it is precisely a given. And certainly there are a number of points and propositions that Manovich assumes and takes for granted in order to construct his analytical framework for the interpretation of large sets of cultural data. Consider, as a very brief example—one Manovich often refers to in lectures and talks and which for all its brevity is nonetheless indicative of the general problem—his use of Cultural Analytics to study the history of art. What Manovich does in this regard is take a set of canonical images illustrative of the development of art over a particular period of time—from "mid-19th century realism, through impressionism, post-impressionism, leading up to early 20th century geometric abstraction"—and automatically extract their different visual qualities by computer.[86] This then enables him to show how the resulting data, arranged into graphs, to all intents and purposes corresponds to the history of art as it is conventionally understood. So as far as the pace of cultural change and revolution is concerned, "around 1870, things are going to get faster, as you have the development from realism to modernism. Then around 1905, the speed ... increases quite dramatically."[87] Yet how interesting is it that Cultural Analytics should more or less confirm the accepted history of art rather than offer a significant challenge to that history or even address it particularly critically?[88] And how surprising is it, given that the study is based on canonical images taken from that same history?[89] Far from enabling him to avoid having to answer the kinds of questions often associated with the close reading by single scholars of a relatively small number of texts and that were dominant for so much of the twentieth century, could Manovich's Cultural Analytics approach to art history not here be said to be based on the assumption that such apparently untimely questions have already been answered—to the extent that they now appear to be relatively unimportant and unproblematic issues, if not indeed a given?

To put things in what are merely the most obvious of terms, what is being understood and brought together as illustrative of the artistic canon? What is left outside, perhaps because it is not perceived as art, or as a canonical image, or does not belong to this particular version of art history? And which paintings are included under the categories of realism, impressionism, postimpressionism, and geometric abstraction? How are we to recognize and understand these different art movements and distinguish them from one another? Are realism, modernism, and so forth defined in the same way everywhere and always? Moreover, which artists are included in the data set? From which countries? How are all these selections and decisions being made? By whom (or what)? With what authority and legitimacy? (And this is before we address technical issues such as those concerned with how accurately the colors, tones, and intensities of a variety of different paintings can be reproduced and compared on a digital screen, let alone across the HIPerSpace display wall of 70-by-30-inch monitors Manovich uses, for these technologies inevitably modify the information they carry.)[90]

Even if Manovich is merely using art history as one of many examples to demonstrate what Cultural Analytics is potentially capable of with regard to democratically broadening the "canon of cultural material under consideration by humanities scholars" and analyzing large sets of cultural data, now and in the future, a number of questions remain.[91] Consider his assertion that today it is perfectly "feasible to computationally analyze all the images contained in all the museums around the world, all feature films ever made, and all the billions of photographs uploaded on Flickr."[92] Indeed, the Cultural Analytics page of the Software Studies Initiative's website describes one of the key goals of Cultural Analytics research as being to "create much more inclusive cultural histories and analysis—ideally taking into account all available cultural objects created in [a] particular cultural area and time period ('art history without names')."[93] Yet what would all the available cultural objects created in a particular cultural area and time period be? What theory of the cultural object—or cultural area and time period, or indeed culture—is being used to underpin such research? And, again, what types of analysis and questions are being privileged, and what assumptions and biases are involved? How are all these images and objects being structured for retrieval and analysis? What is being left out? (At the very least, that would be everything that cannot be so digitized, framed, and structured.) And how do such (non)decisions affect the analysis?[94]

For the most part, rather than taking the time to reflect rigorously on such questions and engage seriously with them, Manovich's Cultural Analytics in effect abstracts the (large sets of) visual cultural objects it chooses to work with—such as the "4535 Time magazine covers, ... 1100 feature films, and one million manga pages" referenced on the Software Studies

Initiative website—from the particular historical, social, and cultural contexts, practices, and sets of relations associated with their production, mediation, interpretation, and consumption (politics, the law, the market economy, and so forth), to focus primarily on the formal aspects of their contents and structure of composition: the color saturation of *Time* magazine covers, for example.[95] Cultural Analytics proceeds to treat these cultural objects and artifacts as if they constitute more or less identifiable, stable, self-identical, some might say essentialist forms that can be analyzed automatically by using "image processing and computer vision techniques" in order "to generate numerical descriptions of their structure and content," thus transforming these (sets of) cultural objects into data.[96] This then allows the Cultural Analytics researcher to perform various "new" kinds of operations and procedures borrowed from computer science and software using these numerical descriptions, such as searching, sorting, copying, combining, comparing, correlating, visualizing, graphing, sharing, and remixing. In doing so, however, Manovich's Cultural Analytics takes too little account of the constitutive force of its own analysis. Just as critical theory tells us that the reader of a text is constituted as a subject in and by the very process of reading, so the (large sets of) objects of Cultural Analytics research do not exist outside and prior to the analysis in any simple or straightforward sense, but are performatively constructed by it (not least through numerous selective, filtering, and hierarchizing procedures) in the very process of being analyzed, translated into data, and operated on, regardless of whether this is done automatically. It is a phenomenon that can variously be understood in terms of the irreducible violence, ambiguity, fictionality—or, following the philosopher and quantum physicist Karen Barad—intra-action that is inherent in all analysis, interpretation, and mediation, the implications of which the past five decades of critical theory have spent a good deal of time endeavoring to understand and think through: hence theory's interest in writing, literature, *poiēsis*, and so on.[97]

Indeed, would it be going too far to suggest that in his desire to develop what he refers to as a "new paradigm for the study, teaching and public presentation" of cultural artifacts, "their dynamics, and flows," Manovich has neglected to pay sufficient attention to taking on and assuming (rather than merely repeating and acting out), the implications of one of the major insights regarding language and technology acquired from twentieth-century theory?[98] It is a lesson the latter has been teaching us since at least the work of Heidegger in "The Question Concerning Technology" (though there are traces as far back as the "first mechanized philosopher" Nietzsche and the development of the typewriter).[99] Moreover, the lesson is by now well known: that it is not just we who speak and act through language and technology; it is also language and technology that speak and act through us.[100] What this means is that we need to ask questions about more than how we can control, search, find, access, order, structure, mine, map, visualize, graph, audit, interpret, analyze, assess, share, and remix vast amounts of cultural data through software tools and techniques and approaches drawn from computer science. We also need to devote great care and attention to asking questions about how these tools, techniques, and approaches are controlling, searching, finding, accessing, ordering, structuring, mining, mapping, visualizing, graphing, auditing, interpreting, analyzing, assessing, sharing, and remixing through, around, and as part of us.[101] And thus we need to explore how they are involved in the process of constituting and organizing our culture and society—and with it, our critical theory and philosophy as well as our sense of the humanities, humanists, and the human—in the twenty-first century.

The argument presented above points to a key problem with the attempt to shift from an interest in the kind of critical theories that dominated the humanities for so much of the twentieth century to an interest in tools, techniques, and methods adapted from computer science and related fields. If we do not explicitly do theory—because we either think we have left it behind or relegated it to some as-yet-unspecified point in the future—we do not end up not doing theory. Every method and methodology contains theory (and this applies even to those that consider themselves to be theory neutral). If we do not explicitly do theory, we merely end up doing simplistic and uninteresting theory that remains blind to the ways it acts as a relay for other forces, including those that are part of the general movement in contemporary society that Lyotard associated with the widespread use of computers and databases and the exteriorization of knowledge. As we have seen, we are experiencing a movement toward business, management, and STEM subjects and away from the humanities; toward a concern to transform knowledge and learning into quantities of information and to legitimate power and control by optimizing the social system's performance in instrumental, functional terms, and away from questions of what is just, right, and true;[102] toward an emphasis on openness, efficiency, and transparency and away not just from a concern with public service and the common good, but also from what is capable of disrupting and disturbing society and what, in remaining resistant to a culture of measurement and calculation, helps maintain much-needed elements of dissensus, dysfunction, ambiguity, conflict, unpredictability, inaccessibility, and inefficiency.

In this respect, there is a temptation to agree with those who have insisted that Manovich's Cultural Analytics is "unconvincing."[103] But could we go further? Could we say his data-driven cultural research functions as an alibi for an unthought-out and rather shallow form of humanities scholarship that has itself been colonized by, and "passionately" imitates, the concerns of scientists, businesses, and government agencies?[104]

> I feel that the ground has been set to start thinking of culture as data (including media content and people's creative and social activities around this content) that can be mined and visualized. In other words, if data analysis, data mining, and visualization have been adopted by scientists, businesses, and government agencies as a new way to generate knowledge, let us apply the same approach to understanding culture.[105]

In taking for granted and following "the templates established by the professionals" and marginalizing positions that go against this emphasis on instrumentality, does such scholarship constitute merely a "further stage in the development of [the] 'culture industry' as analyzed by Theodor Adorno and Max Horkheimer"?[106]

Certainly, as a result of his repeated failure to be rigorously critical and self-reflexive, think long and hard about the consequences of considering computers as analytical tools, and ask larger theoretical questions about contemporary culture and how to make decisions regarding what is just and right, it is often difficult to discern how Manovich is doing too much more in his Cultural Analytics research than augmenting the power and control of states, companies, and multinational corporations by using computers and software to produce, among other things, deliverables that can be marketed and sold, not least in exchange for funding. Yet what makes his Cultural Analytics so fascinating, as I read it, is the way it is clearly striving to open the humanities to some of the new disciplines, frameworks, and forms of knowledge that digital media technologies may make possible. So I conclude this attempt to use Cultural Analytics to think through some aspects of the relation between data-driven scholarship, theory, and critique by taking Manovich at his word and treating his stated interest in cultural criticism, theory, and self-reflexivity seriously. To return to the question with which we began: What forms might the kind of twenty-first-century theory he points us toward—but at the time of this writing he himself apparently is as yet unable to articulate—actually take?

Part 3: Critical Theory in the Twenty-First Century

One starting point for speculating on these questions is provided by the artist, writer, theorist, and fellow participant in the Software Studies Initiative, Eduardo Navas, when he claims that Cultural Analytics, as practiced by Manovich, "is bringing together qualitative and quantitative analysis for the interests of the humanities. In a way Cultural Analytics could be seen as a bridge between specialized fields that in the past have not always communicated well."[107] It is an interpretation that finds support from an account of some of "the promises and challenges of big social data" in which Manovich, displaying more signs of critical reflection than in much of his Cultural Analytics—related research, perhaps comes closest yet to articulating what form such a new cultural criticism might actually take. Here, the study of culture and society throughout the twentieth century is positioned as having relied on two very different kinds of data, "'surface data' about lots of people and 'deep data' about the few individuals or small groups":

> The first approach was used in all disciplines that adapted quantitative methods (i.e., statistical, mathematical, or computational techniques for analyzing data). The relevant fields include quantitative schools of sociology, economics, political science, communication studies, and marketing research.
>
> The second approach was used in humanities fields such as literary studies, art history, film studies, and history. ... The examples of relevant methods are hermeneutics, participant observation, thick description, semiotics, and close reading.[108]

However, Manovich sees the rise of social media in the middle of the first decade of the 2000s, along with computational tools able to handle extremely large data sets, as making possible a "new paradigm" based on a combination of "quantitative and qualitative approaches" (472, 473). Consequently, no longer must we endeavor to chart a third path between these two approaches such as that represented for Manovich by statistics and sampling, enabling researchers to "expand certain types of data about the few into the knowledge about the many," with all the problems attendant on such an expansion (462). Indeed, we do not have to "choose between data size and data depth" at all (462–63). Rather, "'surface is the new depth'" (472) in the sense that

> we can use computers to quickly explore massive visual data sets and then select the objects for closer manual analysis. While computer-assisted examination of massive cultural data sets typically reveals new patterns in this data that even the best manual "close reading" would miss—and of course, even an army of humanists will not

be able to carefully close read massive data sets in the first place—a human is still needed to make sense of these patterns. (468–469)

Encouraged by this line of argument, it is tempting to imagine that all we need to do to resolve the situation facing cultural criticism in the twenty-first century is find a means of marrying the quantitative methods and cultural analysis characteristic of Manovich's research, along with the necessary "expertise in computer science, statistics, and data mining" he sees humanities researchers as typically lacking (470), with the kind of rigorous theoretical critique and self-reflexivity he maintains should also be part of any Cultural Analytics study. We should proceed with care, however. For once embarked on this path, we are likely to find ourselves confronted by a variation on a problem I have detailed elsewhere:[109] that it is not necessarily possible to enhance the performative theoretical interpretations that have long been a prominent feature of the humanities with the kind of positivistic, empirical methods and "tools of quantitative analysis often found in the hard sciences."[110] It is not possible for the simple reason that these different approaches to culture and society do not "complement" each other, as Cohen and Gibbs have it,[111] rather, they remain incommensurable—not least because the dialectical impulse to combine theoretical critique with empirical and quantitative analysis is itself a quite traditional one that theory has in many of its guises worked hard to challenge.[112]

To be clear, this incommensurability does not mean these "specialized fields" are incapable of communicating or interacting, only that they are not able to do so quite as smoothly and straightforwardly as Manovich and others imply. It means they cannot be married, merged, or synthesized, for example;[113] that "human ability to understand and interpret—which computers can't completely match yet—and the computer's ability to analyze massive data sets using algorithms we create" cannot be simply combined.[114] (Indeed, it could be argued that the attempt to produce some kind of joint practice or new synthesis in the humanities by enhancing one with the other risks detracting from the specific advantages and insights of each, while ignoring their antagonisms and incompatibilities.) But what it also means is that any rigorous attempt to think these approaches together needs to begin by explicitly recognizing the incommensurable nature of their relation and thematizing it accordingly. Far more time and care thus need to be spent on how any such communication can be achieved between the respective partners in this impossible relationship than we have seen devoted to it so far.

This is where the lack of rigorous attention on Manovich's part to some of the theories that dominated the humanities for much of the twentieth century is felt most keenly. For certainly Marxism, post-Marxism, psychoanalysis, and deconstruction are in their different ways all capable of providing a potential means not of reconciling the kind of "deep" close reading and self-reflexive theoretical critique that has been so important to the humanities with the "surface," quantitative analysis, and empirical methods more readily associated with the sciences and social sciences, but of producing a consciously developed theory of their incompatibility. Such a theory might even be capable of showing how they can both be practiced at the same time, as two incommensurable positions, in an irresolvable yet productive tension, so that the questions, issues, and approaches specific to each are capable of generating new findings, insights, and realizations in the other—to the point where both of their identities are brought into question. The process of developing such a theory would involve more than merely negotiating the difficult relationship between the two, co-switching emphasis and attention from one to the other and back again, as appropriate. It would not be a case of shifting the epistemological ground so that (in the words of some of those who have also been critical of the computational turn toward data-led methods and have made a case for the continuing importance of the traditional humanities to the digital humanities) the humanities can push back culturally, as well as intellectually, "against the dominant models of a kind of quantitative and empirical approach," and regain some of their confidence in what they do.[115] Nor would it be a matter of performing quantitative statistical modeling and analysis in a less naive and more sophisticated manner than has been carried out by many digital humanists to date, with greater emphasis being placed on modeling conditions and probabilities than on counting things. Nor would it even mean harnessing "digital toolkits in the service of the Humanities' core methodological strengths: attention to complexity, medium specificity, historical context, analytical depth, critique and interpretation."[116] Instead, the development of such a theory would require opening cultural criticism to disciplines, frameworks, and forms of knowledge that are neither close nor distant in their reading practices, neither methodological nor theoretical, quantitative nor qualitative, deep nor surface, digital nor traditional humanities—nor "humanistic," nor "human," for that matter.[117] Rather, they would be, in the words of one twentieth-century commitment to theory, "something else besides"—something that challenges the conventional distinctions between them and, in so doing, "contests the terms and territories of both."[118] It is a theory we might *begin* to think of as being not just postdigital but posthuman and post*human*ities too.[119]

Chapter 3 continues with this exploration of some of the consequences that changes in the media landscape, including those associated with the development of corporate social media such as Twitter, Facebook, and Google+, have for the ways in which critical theorists create, perform, and circulate knowledge and research. Following on from my concern with the humanities and the way in which the quantitative approaches associated with the computational turn in the humanities, far from replacing or supplementing the single human scholar who reads a small number of texts closely, often operate according to quite traditional humanist assumptions themselves, it examines some of the implications any rethinking of theory in the twenty-first century has for an idea on which the humanities—and with it the concept of the university—is based: that of the human subject itself.[120] In order to explore the possibilities for a new model of human subjectivity, it takes as its focus the work of a philosopher described as "our most compelling and ambitious theorist" of posthumanism, Bernard Stiegler.[121] Of particular interest is Stiegler's claim that with the Web and digital reproducibility, we are living in an era in which subjects are created with a different form of the awareness of time. However, in line with both the above argument and Stiegler's own insistence that a new critique of political economy needs to be developed that is capable of responding to an epistemic environment very different from that known by Marx and Engels, chapter 3 refuses to simply contrast print and the Web in terms of an offline-online dialectic. Instead, it proceeds by paying special attention to the medium Stiegler himself employs most frequently to analyze the relation of subjectivity, technology, and time: the linearly written and organized, print-on-paper codex text, with all its associated concepts, values, and habitual practices.

3 THE HUMAN
#MySubjectivation

Over the past few years, a number of critical theorists and radical philosophers have positioned digital media information technologies, and corporate social media and social networks in particular, as contributing to the development of a new kind of human subjectivity. It is a subjectivity suffering from attention deficit disorders that is rendered anxious, panicked, and deeply depressed by the accelerated, overstimulated, overconnected nature of life under postindustrial capitalism in the twenty-first century.[1] Others have been keen to portray the Tahrir Square, Taksim Gezi Park, and Yo Soy 132 protests in Cairo, Istanbul, and Mexico, together with the more recent demonstrations in Baltimore and Brazil , as expressive of new ways of being human that are markedly different from those generated by neoliberalism.[2] My question here is whether, in the era of Anonymous, the *indignados*, and Occupy, with their explicit rejection of the drive toward individual fame that constitutes an inherent part of modern capitalist society, and emphasis instead on nonhierarchical forms of organization, we need to explore new ways of being radical theorists and philosophers too.[3] Are ways of being and doing now required that are unlike us, at least as we currently live, work, act, and think, in that they are not so tightly bound up with the logic of the contemporary cultural industries?

Significantly, few of the key theorists whose thought provides a framework for the study of the relationship between culture and society have paid serious attention to the implications that changes in the media landscape have for their own ways of creating, publishing, and circulating knowledge and research (and this is the case despite the opportunities that are provided by networked digital information machines especially to perform ideas of the human, the author, the text, the book, originality, copyright, and so on differently).[4] The majority have been content to operate with norms, conventions, practices, and modes of production that originated in very different eras. Indeed, a number of them would be familiar to scholars in the second half of the seventeenth century, when the world's first peer-reviewed journal was established, let alone the nineteenth or twentieth.[5] With surprisingly rare exceptions, they are those of the rational, liberal, humanist author working alone in a study or office. Motivated by a "desire for pre-eminence, authority and disciplinary power," to quote Stanley Fish's characterization of his own ambition as a literary critic, this author produces a written text designed to make an argument so forceful and masterly it is difficult for others not to concur.[6] Claiming it as the original creative expression of his/her own unique mind, the lone author submits the written work for publication as part of a paper (or paper-centric) journal or book. Once the work has been peer reviewed and accepted for publication, it is eventually made available for sale under the terms of a publisher's policy, license, or copyright agreement. The latter asserts his/her right to be identified and acknowledged as its author and to have it attributed to him/her as his/her intellectual property; transfers the rights to the commercial exploitation of the text or work as a commodity that can be bought and sold to the publisher; reserves the right to control and determine who publishes, circulates, and reproduces the text, how, where, and in which contexts; and prevents the integrity of the original, fixed, and final form of the text from being modified or distorted by others.

Yet if the majority of theorists have remained somewhat blind to the implications of changes in the media landscape for their own ways of performing knowledge (a landscape that shapes even if it does not determine human consciousness), one important thinker who has paid sustained attention to the relation between subjectivity, technology, and time is the French philosopher Bernard Stiegler. It is to Stiegler's work that I now turn for help in order to think through still further the relation between networked digital media technologies, temporality, and our ways of living, working, acting, and thinking as critical theorists.

That said, it is impossible to provide a full account of Stiegler's oeuvre here, such is its size and scope. The back cover blurb of the 2009 English translation of *Technics and Time, 2: Disorientation,* refers to his having published seventeen books in "the last five years alone." What are we to make of this extreme (over)productivity on his part? Is it in its own way an instance of the speed Stiegler links to disorientation and the industrialization of memory in this book? This possibility haunts much of what follows, which, aware a great deal of work is still to be done on the many issues that Stiegler's philosophy raises, constitutes merely an initial attempt to contribute to any such future study.

The Philosophical Impossibility of Unliking Media Technologies in the Mind of Someone Living

Building on the philosophy of Jacques Derrida, Stiegler argues that the relation of the human to technology is one of originary technicity or prostheticity. What this means is that, contrary to the classical Aristotelian view, technology (i.e., that which is organized but inorganic, manufactured, artificial) is not added to the human from the outside and only after the latter's birth, as an external prosthesis, tool, or instrument used to bring about certain ends. Rather, the human is born out of its relation to technology. As far as Derrida is concerned, the association of time with the technology of writing means that this originary relation between technology, time, and the human can be understood as a form of writing, or arche-writing (i.e., writing in general, which is "*invoked* by the themes of 'the arbitrariness of the sign' and of difference"—as opposed to any actual historical system of writing, including that of speech).[7] As Stiegler asserts in a relatively early essay, "Derrida and Technology," all media for Derrida, "beginning with the most primal traces ... and extending as far as the Web and all forms of technical archiving and high-fidelity recording, including those of the biotechnologies ... are figures, in their singularity, of the originary default of origin that arche-writing constitutes."[8] For Stiegler, however, such an understanding universalizes arche-writing and underplays the specificity of different media technologies and their relation to time. Instead, he emphasizes the historical and contingent nature of this relation. Put simply, because the human is born out of a relation to technology, and because time is possible and can be accessed and experienced only as a result of its prior inscription in concrete, technical forms, the nature of subjectivity and consciousness changes over time as media technologies change. Drawing on the argument of the paleontologist André Leroi-Gourhan, to the effect that the emergence of the human species coincided with the use of tools, Stiegler presents this process as having begun in the Upper Paleolithic period, its most recent stage being the Web. In "The Discrete Image," another early essay, this one on the epistemology of digital photography, he thus stresses that we must distinguish between:

- the reproducibility of the letter, first handwritten and then printed;
- analog reproducibility (i.e., photographic and cinematographic), which Benjamin studied extensively;
- digital reproducibility.

It is "these three great types of reproducibility," Stiegler insists, that "have constituted and overdetermined the great *epochs* of memory" in the West, producing eras in which subjects are created with different forms of the awareness of time.[9]

At this point a similar criticism can be made of Stiegler—and by implication of those theorists of digital media who have followed him in this respect, such as Mark Hansen and N. Katherine Hayles, whose positions build on Stiegler's use of the related concept of technogenesis—as he makes of Derrida.[10] Just as Derrida sees all media as figures of the originary default of origin constituted by arche-writing, Stiegler himself argues "for a generalised technicity—especially as a condition of temporality."[11] From a more strictly Derridean viewpoint, then, Stiegler does not do enough "to preserve the ontological difference between the technical synthesis of time and *différance* as the quasi-transcendental condition of possibility for time."[12] Nevertheless, despite this (and in a sense precisely because of it), Stiegler's work can be extremely helpful when it comes to thinking through the role that the changing technical environment, and with it the emergence of digital media technologies, plays in the production of human subjectivity. This can be demonstrated by turning to his understanding of the cultural industries.

To simplify his argument for the sake of economy, Stiegler presents the cultural industries as subordinating the subject's consciousness and experience of time to the formalized, standardized, reproducible, and controllable routines of their "temporal industrial objects." The cultural industries, and particularly the program (radio and television) industries within them, achieve this by connecting people and their attention to the same regular radio programs, TV broadcasts, and so forth on a mass basis. Accordingly, there is too little scope for the event, for singularity—for the "welcoming of the new and opening of the undetermined to the improbable," to play on his "idea of value defined as knowledge" from *Technics and Time, 2*.[13] Newspapers, for example, he describes as being machines "for the production of ready-made ideas, for 'clichés,'" motivated by the demands of short-term profit, whose "criteria of selection are aspects of marketability."[14] As a consequence, the cultural and program industries interfere with the ability of each subject to singularly appropriate and transform what Stiegler, following Gilbert Simondon, calls the *preindividual fund*, which is the process that results in the psychic individuation of each individual—so much so that in a more recent essay, Stiegler is able to show how they function to suffocate desire and destroy the individual:

As heritage of the accumulated experience of previous generations, this preindividual fund exists only to the extent that it is singularly appropriated and thus transformed through the participation of psychic individuals who share this fund in common. However, it is only shared inasmuch as it is each time *individuated,* and it is individuated to the extent that it is *singularised.* The social group is constituted as *composition* of a synchrony inasmuch as it is recognised in a common heritage, and as a diachrony inasmuch as it makes possible and legitimises the singular appropriation of the pre-individual fund by each member of the group.

The program industries tend on the contrary to *oppose* synchrony and diachrony in order to bring about a hyper-synchronisation constituted by the programs, which makes the *singular* appropriation of the pre-individual fund impossible. The program schedule replaces that which André Leroi-Gourhan called socio-ethnic programs: the schedule is conceived so that my lived past tends to become the same as that of my neighbours, and that our behaviour becomes herd-like.[15]

Perhaps one of the most important things to be learned from Stiegler is that the way to respond responsibly to this "industrialization of memory" and the threat it poses to the intellectual, affective, and aesthetic capacities of millions of people today, is not by trying to somehow escape or elude the technologies of reproduction, or become otherwise autonomous from them. Originary technicity means there is no human without technology, as the "*who* is nothing without the *what,* since they are in a *transductive* relation during the process of exteriorization that characterizes life."[16] Any such response must itself therefore involve these technologies. By the same token, neither can we proceed in the hope that the mass media of the cultural and program industries are eventually going to disappear or be abolished, or that we can replace them and the alienating affects of their one-to-many broadcasting model with the apparently more personal, participatory, many-to-many (as well as many-to-one and one-to-one) model associated with the dominant networked digital media technologies. Witness the way a relatively small number of powerful corporations, including Amazon, Apple, Facebook, and Google, are currently in the process of supplementing, if not entirely superseding, the "old" cultural and program industries with regard to the subordination of consciousness and attention to preprogrammed patterns of information conceived as merchandise. They are doing so by exposing users to cultural and cognitive persuasion and manipulation (often but not always in the form of advertising) based on the tracking and aggregation of their freely provided labor, content, and public, personal, and embodied data. This process is aimed at targeting individual users on a finely grained, personalized, and, with mobile media, even location-sensitive real-time basis: "If a bloke is walking through Manchester and checks the football scores on his phone, he could see an ad for a discount voucher at M&S [Marks and Spencer], which is

300m from where he's walking. … A mobile signal indicates where a user is situated within 15 metres."[17]

Stiegler presents such technologies as *hypomnémata*: forms of *mnemonics* (cultural memory) that Plato described as *pharmaka*, or substances that function, undecidably, as neither simply poisons nor cures. Rather than reject or critique them outright, he suggests we need to explore how some of the tendencies of which our current economy of the *pharmakon* is composed can be deployed to give these technologies new and different inflections. As he posits in a 2009 book arguing for the development of a new critique of political economy as "the task *par excellence* for philosophy" today, this "economy of the *pharmaka* is a therapeutic that does not result in a hypostasis opposing poison and remedy: the economy of the *pharmakon* is a *composition* of tendencies, and not a dialectical struggle between *opposites*."[18]

Of course, variations on the idea that reproductive media technologies—including corporate (i.e., privately owned) social media and social networks such as Twitter and Facebook—are neither simply "good or bad, productive or distracting, enabling or dangerous," have been put forward a number of times.[19] With more and more people today accessing the Internet using tethered mobile devices—smart phones, tablets, and e-readers—controlled by their manufacturers, those that provide their operating systems, or the telecommunications companies that run the mobile networks, some critics propose radically *unliking* private, closed, and semiclosed systems, including those represented by Apple's single-purpose apps, iDevices and iCloud computing. They advocate time and attention be given instead to those tendencies within our current economy that encourage physical infrastructure and networks that are less centralized and more open to being continually updated, interrupted, reappropriated, transformed, and reimagined. The emphasis is on infrastructure and networks that make it easier for users to understand how such media are made, "in order to restart the contract on different terms" and give users "the right of response, right of selection, right of interception, right of intervention," to draw on Stiegler's televised conversation with Derrida.[20] The latter tendencies manifest themselves in the phenomena of much so-called Internet piracy, the "hacktivism" associated with 4chan and Anonymous, as well as in "alternative free and open source software that can be locally installed" by a range of different groups dedicated to working together to get things done, thus generating a "multitude of decentralized social networks … that aspire to facilitate users with greater power to define for themselves with whom [to] share their data."[21] (It is harder for free software, the code of which is open to everyone to inspect and improve, to contain the kind of secret back doors that enable corporations and government intelligence agencies to gather data about users, for example.) Numerous such collaborative alternatives to the dominant social media and social networking monopolies are available, although for obvious reasons they are less well known than their corporate counterparts. FreedomBox, for instance, is "a community project to develop, design and promote personal servers running free software for distributed social networking, email and audio/video communications"; Diaspora is also a federated social network that can be used instead of Facebook; unCloud is an art project-cum-software-application that "enables anyone with a laptop to create an open wireless network and distribute their own information"; and Etherium is a platform and a programming language that enables developers to "build and publish next-generation distributed applications" on a decentralized network.[22]

Yet when it comes to considering the relation between digital media technologies and our ways of being, thinking, and doing as theorists and philosophers, a more intriguing question, I suggest, is one that often remains overlooked, or otherwise ignored in academic discussions of YouTube, Instagram, LinkedIn, Tumblr, Snapchat, WhatsApp, Crabgrass, Lorea, and so on. This question concerns the very medium Stiegler himself employs most frequently and consistently to critique the specific changes in technology that are helping to shape subjectivity in the era of digital reproducibility: the linearly organized, bound, and printed paper codex text. How appropriate is it for Stiegler to analyze and critique such changes as if he himself were in the main living and working in the epoch of writing and the printed letter, with all that implies with regard to his way of being as a philosopher? Is Stiegler—like Derrida before him, on his account—not in his own fashion privileging writing, and the associated forms and techniques of presentation, debate, critical attention, observation, and intervention, as a means of understanding the specificity of networked digital media technologies and their relation to cultural memory, time, and the production of human subjectivity?[23]

Stiegler's notion of originary technicity and the default of origin undermines the romantic, possessive, humanist conception of the self as separate from those objects and technologies that provide it with a means of expression: writing, the book, film, photography, the Web, smart phone, tablet, and so forth.[24] Yet from the very first volume of *Technics and Time* (originally published in French in 1994) through to the 2014 appearance in English of *The Lost Spirit of Capitalism: Disbelief and Discredit*, volume 3, and beyond, Stiegler to all intents and purposes continues to act as if he genuinely subscribes to the notion of the author as individual creative genius associated with the cultural tradition of European romanticism.[25] He persists in continually publishing books, including a number of multivolume monographs, devoted to the building of long-form "arguments that are intended to be decisive, comprehensive, monumental, definitive" and, above all, his.[26] In *Acting Out*, for example—which, interestingly, is composed of two short books on how he became a philosopher and narcissism, respectively—Stiegler repeatedly uses phrases such as this is what "I call 'primordial narcissism ... the 'becoming-diabolical' ... a tertiary retention ... hypersynchronization."[27] Indeed, at least in their compulsive repetition of the traditional, preprogrammed, ready-made methods of composition, accreditation, publication, and dissemination (not to mention layout, design, reliance on embodied means of navigation, and so on), his books very much endeavor to remain the original creation of a stable, centered, indivisible, and individualized, humanist, proprietary subject.

Of course, it is not only Stiegler who acts out what it means to be a critical theorist or radical philosopher by writing and publishing in this fashion. Much the same can be said of Catherine Malabou, Chantal Mouffe, Jacques Rancière, Slavoj Žižek—in fact, most thinkers of contemporary culture, media, and society today.[28] This point even applies to those theorists of digital media who know how to code and produce experimental e-literature, such as Alexander R. Galloway, Wendy H. K. Chung, and N. Katherine Hayles. How can it be otherwise when academics in the humanities need at least one monograph published with a reputable print press to secure that all-important first position or tenure (and that is after having produced a 60,000- to 80,000-word PhD thesis consisting of "original," discrete work, of which they have to officially declare themselves the sole "author")?[29] Don't we all acquire much of our authority as scholars by acting romantically as if we were still living in the epoch of writing and print? Would we attach the importance to Stiegler's work we do if he had not (single-)authored so many codex books? Would Stiegler still be considered a serious thinker and philosopher, and would most of us even have heard of him, had he operated in less conventional academic terms, merely as part of the Ars Industrialis association of cultural activists he formed in 2005, or any of the institutes he is connected to?[30] The latter include not just the Centre Pompidou's Innovation and Research Institute (IRI), which he directs, but also the INA (Audiovisual National Institute), where Stiegler moved the research department toward signal processing and analysis, and IRCAM (Institute for Acoustic and Musical Research Coordination), where he did something similar in the field of sound.[31] As he put it an interview at the 2012 International World Wide Web Conference when discussing his relationship to some of these projects:

The new dynamics of knowledge needs henceforth that Web issues be questioned, practiced, theorized and critically problematized ... as with the Bologna University during the 11th century, then with the Renaissance era, then with the Enlightenment and Kant's question in *Le conflit des facultés*, we are living a significant organological change—knowledge instruments are changing and these instruments are not just means but rather shape an epistemic environment, an episteme, as Michel Foucault used to say.[32]

Nevertheless, for all his activities with IRI, INA, IRCAM, Ars Industrialis, and now Pharmakon.fr to develop a new, enlarged organology for the contemporary era that includes digital technology, networks, and software, the question remains:[33] if Stiegler is right, and with the Web and digital reproducibility we are now living in an era in which subjects are created with a different form of the awareness of time, to what extent can this episteme and the associated changes in the media ecology that are shaping our memories and consciousness be understood, analyzed, rethought, and reinflected by subjectivities that, to a significant extent, continue to live, work, and think on the basis of knowledge instruments originating in a very different epistemic environment?

Capital as Academic Subjectivation Machine

To explore this question and its implications for critical thinkers further, let us return to Stiegler's claim that the task par excellence for philosophy now is the development of a new critique of political economy capable of responding to an epistemic environment very different from that known by Marx and Engels.[34] Stiegler has been held up by software theorist Alexander Galloway as "one of the few people writing today" who approaches Gilles Deleuze's idea of the control society seriously both "as a political and philosophical problem" and as a critique of political economy.[35]

As is well known, not least as a result of important work by Michael Hardt, Antonio Negri, and Maurizio Lazzarato, Deleuze's thesis is that we are no longer subject primarily to those closed, disciplinary modes of power Michel Foucault traced historically in *Discipline and Punish* and *The Birth of the Clinic*. These govern by means of a dispersed and decentralized ensemble of institutions, instruments, techniques, procedures, and apparatuses (or *dispositifs*) that operate to produce and regulate subjectivity—customs, habits, thoughts, behavior, and so on—via the interiorization of the law. What Deleuze refers to as disciplinary societies are characterized by vast closed environments—the family, school, barracks, factory, and, depending on circumstances, the hospital—each with its own laws, through which the individual ceaselessly passes. Disciplinary environments or enclaves are about enclosure, confinement, and surveillance: their project is to "concentrate," "distribute in space," "order in time," "organize production," "administer life," "compose a productive force within the dimension of space-time whose effect will be greater than the sum of its component forces."[36] Above all, however, it is the prison that serves as the "analogical model" for the closed system of disciplinary societies and the manner in which it produces and organizes subjectivity—hence Foucault's question in *Discipline and Punish*: "Is it surprising that prisons resemble factories, schools, barracks, hospitals, which all resemble prisons?"[37]

According to Deleuze, such disciplinary societies reached their peak at the beginning of the twentieth century. His contention is that just as Foucault saw modern disciplinary societies as having superseded the ancien régime's "societies of sovereignty" from the late eighteenth century onward, so, in a process that has accelerated after World War II, social organization is ceasing to be disciplinary, if it has not done so already—to an extent that all the closed spaces associated with disciplinary societies (the family, the health service, the factory system) are in now crisis. These disciplinary societies are in the process of being replaced by societies of control. The latter are our "immediate future," Deleuze tells us, and contain extremely rapid, free-floating forms of "continuous control and instant communication" that are interiorized within the minds and bodies of subjects themselves and operate in environments and spaces that are more fluid and open, extending well beyond social institutions such as the family and factory.[38]

An example from the twenty-first century, by now already familiar, comes in the form of enhancements in computer processing capacity and the associated availability of large, complex data sets—sometimes, not always correctly, called big data.[39] These are enabling a degree of data mining and pattern recognition to be achieved that makes it possible for algorithms to automatically anticipate and predict—and thus control, albeit in a comparatively open, flexible fashion that is immanent to both the social field and to subjectivity—actions and decisions on the part of the subject before they actually take place. So we have social networks such as Last.fm employing scrobbling software to detail the listening habits of its users and provide them with personalized selections of music based on their previous listening history. Tastebuds.fm even arranges dates for people based on Last.fm scrobbles. To continue with the relationship theme, experts can use

Facebook to tell when two people may be about to get together, as the frequency with which they check out each other's pages increases; Uber knows from its data harvesting when a RoGer, or Ride of Glory, has taken place in one of its taxis after such a hook-up; and Target is able to discern from the data gathered from its stores which shoppers are pregnant, and even when they are going to give birth.[40] Toward the other end of the relationship arc, Marissa Mayer, ex-Google vice president and now CEO of Yahoo! goes so far as to claim that "credit card companies can predict with 98% accuracy, two years in advance, when a couple is going to divorce, based on spending patterns alone."[41]

Instead of the prison or factory of disciplinary societies, then, what we have is the corporation of the control societies, which Deleuze likens to a spirit or gas. And one of the main differences between the factory and the corporation is this:

> The factory constituted individuals as a single body to the double advantage of the boss who surveyed each element within the mass and the unions who mobilized a mass resistance; but the corporation constantly presents the brashest rivalry as a healthy form of emulation, an excellent motivational force that opposes individuals against one another [when it comes to negotiating for a higher salary, for example, according to the modulating principle of individual performance and merit] and runs through each, dividing each within.[42]

It is a state of affairs that pertains not just to corporations or social media and social networks. As Deleuze makes clear, it also applies to other areas of society such as the school. It is important to be aware of this because, for Stiegler, education, beginning with that which the mother bestows on her child, is the attention-forming modality through which the process of psychic and collective individuation is concretized.[43] Yet even in the school, the micromanagement of opposed individuals now rules by means of the introduction of an audit culture, evaluation forms, performance targets, league tables, and other forms of measuring and monitoring both students and teachers, with continuous control—including constant assessment, training, staff development, and performance review—replacing the test and examination. Indeed, it is a process of control that is all permeating and never ending, for nothing is left alone for long in a control-based system.[44] While "in the disciplinary societies one was always starting again," as the individual moved from school, to the military, to the factory, in societies of control, one can never finish anything, "the corporation, the educational system, the armed services being metastable states coexisting in one and the same modulation, like a universal system of deformation."[45]

If Deleuze's control society thesis is accepted, it follows that the contemporary university is not best thought of as a factory—despite what some of the slogans of those protesting against the marketization of the higher education system and increase in tuition fees in England proclaim.[46] Nor, to take just one more of the examples I gave earlier, is Facebook, for all the latter's harnessing of the free labor power generated by social cooperation.[47] (It is reported to have enrolled over a seventh of the world's population.) Facebook's flexible, fluid, and relatively open environment, together with its own origins (like Google) in the contemporary university (Facebook was famously invented by a Harvard undergraduate, Mark Zuckerberg), means that it too is far closer to Deleuze's account of the gas-like corporation that has replaced the factory in a control society. And, like the university (and open access, open science, and open data on this reading—see chapter 2), Facebook can be seen as part of the corresponding reconfiguration of the individual in terms of the "dividual" and of the mass in terms of coded data that is produced to be controlled:

> The disciplinary societies have two poles: the signature that designates the *individual*, and the number or administrative numeration that indicates his or her position within a *mass*. This is because the disciplines never saw any incompatibility between these two, and because at the same time power individualizes and masses together, that is, constitutes those over whom it exercises power into a body and molds the individuality of each member of that body. ... In the societies of control, on the other hand, what is important is no longer either a signature or a number, but a code: the code is a *password*. ... The numerical language of control is made of codes that mark access to information, or reject it. We no longer find ourselves dealing with the mass/individual pair. Individuals have become "*dividuals*," and masses, samples, data, markets, or "*banks*."[48]

Now, as I have said, Stiegler has been positioned as someone who takes Deleuze's idea of the control society extremely seriously, as both a political and philosophical issue *and* as a critique of political economy. For him,

> the great delusion is no longer the "leisure society," but the "personalisation" of individual needs. ...
>
> Through the identification of users (user profiling) and other new methods of control these cognitive technologies allow for the subtle use of conditioning, invoking Pavlov as much as Freud. For example, services that encourage readers of one book to read other books read by readers of the same book. Or those internet search engines that promote the most consulted references, thus at once multiplying their consultation and constituting an extremely refined form of viewer rating.[49]

Yet in one respect at least, the control society is something Stiegler—in common with the majority of theorists who have alerted us to the power of algorithms—does not take anywhere near seriously enough. For if "the *what* invents the *who* just as much as it is invented by it"[50]—if, in Galloway's words, "one must today focus special attention on the way control acts on the realm of the 'immaterial': knowledge work, thought, information and software, networks, technical memory, ideology, the mind," in order to follow Stiegler in shifting "from a philosophy of 'what is' [being, ontology] to a philosophy of 'what does'" (what affects, what cares, which is a question of practice, ethics, politics)—then taking Deleuze's idea seriously as a critique of political economy must surely involve paying careful critical attention to our own modes of production and ways of acting and thinking as theorists and philosophers.[51] In other words, we need to consider seriously how the economy of control invents us and our own knowledge work, philosophy, and minds, as much as we invent it, by virtue of the way it modifies and homogenizes our thought and behavior through its media technologies.[52]

What is particularly interesting about Deleuze's thesis from this perspective is that it is not just the prison, factory, or school of the disciplinary societies that is identified as being handed over to the corporation of the control societies. So is the institution in which many theorists and philosophers actually work and think, namely, the university. To draw on the contemporary UK context for a moment, the fundamental transformation in how universities in England are viewed, which was proposed by the New Labour government–commissioned Browne Report published in 2010, and imposed by the subsequent Conservative/Liberal Democratic coalition, albeit with some modifications designed to generate further competition between institutions, such as the introduction of a free market for students with A-level (a high-school final exam) grades of AAB/ABB upward, provides perhaps the highest-profile evidence of this state of affairs.[53] It entails a shift from perceiving the university as a public good financed mainly from public funds, to treating it as a "lightly regulated market." Consumer demand, in the form of the choices of individual students over where and what to study, here reigns supreme when it comes to determining where the funding goes, and thus what is offered by competing "service providers (i.e., universities)," which are required to operate as businesses in order "to meet business needs."[54]

The consequences of handing the university over to the corporation are far from restricted to a transformation in how the university is viewed as an institution, or even to the production of the student as consumer. This process is also having a profound impact on us as academics and scholars (i.e., on that part of what some radical philosophers call the cognitiarian class that actually includes those philosophers themselves). Thanks to the Research Assessment Exercise (RAE) and its successor, the Research Excellence Framework (REF, which is a system for assessing the quality of research in higher education institutions in the UK), many university professors are now given lighter teaching loads and even sabbaticals to allow them to concentrate on their research and achieve the higher ratings that will lead to increases in research profile and the generation of income for their institutions from government, businesses, and funding agencies. Individuals successful in doing so are then rewarded with even more funding and sabbaticals, which only increases the gap between these professors and those who are asked to carry a greater share of the teaching and administrative load. One result is the development of a transfer market—and even a transfer season as the deadline for the next REF approaches—whereby research stars are enticed to switch institutions by the offer of increased salaries, resources, support, and status. At the same time, the emergence of more corporate forms of leadership, with many university managers now being drawn from the world of business rather than the ranks of academe, has resulted in a loss of power and influence on the part of professors over the running of their institutions, for all they may be in demand for their research and publications. Many institutions currently require commercial (rather than purely intellectual) leadership from their professoriate, in line with the neoliberal philosophy that society's future success and prosperity rests on the corporate sector's ability to apply and exploit the knowledge and innovation developed in universities.

Professors and others in leadership roles are not the only ones affected. Most academics today belong to a "self-disciplining, self-managed form of labor force"—one that "works harder, longer, and often for less [or even no] pay precisely because of its attachment to some degree of personal fulfillment in forms of work engaged in."[55] This is in part a result of their having—or indeed wanting, or at least wanting to be seen as wanting (for reasons to do with anxiety over the security of a job they find fulfilling to some extent and because it is their individual entrepreneurial personalities that academics are competitively promoting, marketing, and selling)—to take on greater and intensified teaching and administrative loads, due to severe reductions in government spending on universities combined with an expansion in student numbers, along with the privileging of research stars. The increase in the number of nontenured and non–tenure track, fixed-term, part-time, hourly paid, zero-hour, temporary, and other forms of contingent positions (instructors, teaching assistants, post-docs, unpaid "honorary" research assistants) as we enter more deeply into a precarious labor regime is another significant aspect of the changing higher education environment. The result is a process of casualization and proletarianization that Stiegler has described in a broader context as a loosing of knowledge, of *savor*, of existence, of "*what takes work beyond mere employment*," and as thus leading to a short-circuiting of individuation.[56] Yet academics are also working longer and harder (and faster) as a consequence of the increasing pressure to be constantly connected and prepared for the real-time interaction that is enabled by laptops, tablets, smart phones, apps, e-mail, SMS, Dropbox, and Google Docs. Mobile media and the cloud mean scholars can now be found at work—checking their inbox, texting, chatting, blogging, retweeting, and taking part in online classes, discussions, and forums—not just in their office or even on campus, but also at home, when walking in the city, traveling by train, or waiting at an airport in a completely different time zone from the rest of their institution. The pressure that is created by various forms of monitoring and measurement for academics to show they are always on and available by virtue of their prompt responses to contact from colleagues and students exacerbates this culture of "voluntary" self-surveillance and self-discipline. So too does the increasing use of electronic diaries open to scrutiny, together with swipe card readers that provide university management with data on where faculty are at any given time. As a result, it is becoming harder and harder for academics to escape from (the time of) work.[57] (Building on the notion of the "quantified self," could we call this development the *quantified academic*?)[58]

If the university, like the school, is "becoming less and less a closed site differentiated from the workspace as another closed site," the same can be said of another important aspect of how the control economy and its media technologies are inventing us and our own knowledge work, philosophy, and minds: academic publishing.[59] This can also be seen as undergoing a process of transition: from the walled, disciplinary gardens represented by scholarly associations, learned societies, and university presses to more open, fluid environments. Consider the emphasis currently placed by governments, funding agencies, and institutional managers on the more rapid, efficient, and competitive means of publishing and circulating academic work they associate with the movement for open access. Publishing research and data on such an open basis is heralded as being beneficial by these key players, as it facilitates the production of journal- and article-level metrics for national research assessment exercises, international league tables, and other forms of continuous control through auditing, monitoring, and measuring processes (including the REF, the panels of which now include members drawn from the business community). It also helps to expand existing markets and generate new markets and services. (Tools for metrics and citation indexes are frequently owned by corporations, as in the case of Thomson Reuters' Web of Science and Elsevier's Scopus.) In this respect, the push for open access and open data can be said to dovetail all too seamlessly with the neoliberal philosophy that assigns universities the task of carrying out the basic research the private sector has neither the time, money, nor inclination to conduct for itself, while nonetheless granting the latter access to that research and the associated data to enable their commercial application and exploitation.[60]

Further evidence of a shift in academic publishing toward the kind of open and dispersed spaces associated with Deleuze's thesis is provided by the large number of researchers who are currently taking advantage of the opportunities to acquire authority and increase the size of their academic footprint that are offered by the dominant corporate social media and social networks. As with other areas of the control economy, social networks such as Academia.edu, Facebook, Google+, and LinkedIn are characterized by what Beverley Skeggs calls a "compulsory individuality."[61] Thanks not least to their entry procedures and use of real name policies, the only way to publicly join in with and become part of the communities generated by many of these corporate networks is through one's own personal (self-)profile. (Facebook's terms of service, for example, include the stipulations, "You will not provide any false personal information on Facebook, or create an account for anyone other than yourself without permission," and, "You will not create more than one personal account"; and until recently, you could not use multiple identities or pseudonyms on Google+ unless you were a prominent public figure known by such a pseudonym.)[62] By taking responsibility on themselves for managing, branding, promoting, and marketing their work, ideas, and charismatic authorial personalities in this way using these communication platforms (even if they are accompanied, as they often are, by an attempt to self-consciously deny or downplay any overt desire for individual attention and recognition),[63] academics can be seen to be caught in modern capital's subjectivation machine just as much as the workers "Bifo" and Maurizio Lazzarato describe:

> Capitalization is one of the techniques that must contribute to the worker's transformation into "human capital." The latter is then personally responsible for the education and development, growth, accumulation, improvement and valorization of the "self" in its capacity as "capital." This is achieved by managing all its relationships,

choices, behaviours according to the logic of a costs/investment ratio and in line with the law of supply and demand. Capitalization must help to turn the worker into "a kind of permanent, multipurpose business." The worker is an entrepreneur and entrepreneur of her/himself, "being her/his own capital, being her/his own producer, being her/his own source of revenue" (Foucault). ...

This idea of the individual as an entrepreneur of her/himself is the culmination of capital as a machine of subjectivation.[64]

Publishing today is consequently not an activity academics take part in just for and at work: with as many as a third of scholars reported to be on Twitter, they publish, and act as entrepreneurs and entrepreneurs of themselves and their own subjectivities, in all aspects of their life, in all their "relationships, choices, behaviors."[65] (Actually many social media communities informally encourage this. It is difficult to acquire high status and authority on Twitter, for example, merely by sending tweets that promote you and your work. You have to share more personal aspects of your life and "what's happening?" with you. Only by constructing this carefully curated "self" is it possible to gain the trust and respect of the community to the extent that a small proportion of work-related tweets are socially acceptable.) The separation between work and nonwork is thus becoming difficult for many academics to maintain. Is it work, leisure, or play when you are monitoring Twitter steams, writing an entry on your WordPress blog, adding a bookmark to Delicious, tagging a photograph on Pinterest, or detailing your "likes" on Facebook regarding the books you read? Even if these are forms of leisure, are they ways of spending free time or of controlling it?

Forgetting Stiegler

If Deleuze's idea of the control society is to be taken seriously as a critique of political economy and power relations between the social and the technical, then, as Stiegler suggests it is (although, as we shall see, a question mark can be placed against just how seriously Stiegler actually takes this critique himself), it clearly has significant implications for academic work. The manner in which the latter is increasingly being formed, organized, categorized, stored, managed, published, disseminated, marketed, and promoted now appears very much as a means by which the attention of academics is captured and their thought and behavior modified, homogenized, and sold to entrepreneurs, venture capitalists, shareholders, and advertisers, along with governments, university managers, and funding agencies. (The message is that you need to join everyone else and do this;

you need to publish on an open access, open data basis, and contribute to the upsurge of user-generated content on websites, apps, and social and mobile media if you want to be up-to-date, keep in touch with what's happening, network, build your career, increase your readership and citations, have impact.) Many of today's university workers are thus left with little time in which they are able to direct their attention free from these forms of control.

Faced by this situation, some scholars and academics have looked back to the values of the traditional university as offering an alternative to the *becoming business* of the contemporary institution.[66] In this context, can the continuing maintenance of the values that are associated with writing and publishing print-on-paper codex texts be regarded as having a similarly alternate, oppositional, counter, or radical aspect—for all they can take on a somewhat reactionary appearance in an era in which digital reproducibility, according to Stiegler, constitutes and overdetermines the relation between human subjectivity, memory, and time? To put this question explicitly in the language of Stiegler's philosophy, if "technical development is a violent disruption of extant programs that through redoubling give birth to a new programmatics" (he provides as an example the expansion of orthographic writing in classical Greece) and if this is something that is itself "a process of psychic and collective individuation" ("contemporary disorientation" being the "experience of an incapacity" to bring about such an "epochal redoubling," according to Stiegler), can the writing and publishing of papercentric articles, monographs, and multivolume series of books today help to program the epochal redoubling of our current technical system of reproduction so as to produce just such a new programmatics, thus countering the tendency to subjectivation and disindividuation of the economy of control and its cultural and program(ing) industries?[67]

The wish to sustain a discerning critical understanding and analysis of the specificity of digital media technologies certainly goes a long way toward accounting for Stiegler's own continuing substantial investment in writing and the printed letter as both a medium and material practice, along with the associated forms and techniques of presentation, debate, critical attention, and intervention. After all, "critical thought or reflection" as far as he is concerned is a "fundamental product of the paradoxical double dimension of memory that appears with linear writing" (i.e., the process of textual identification that enables both the grammatical rules of the production of texts to be identified and for "the endurance of their most fundamental irregularity, whose interpretability is the sign", and which is "thus a test for reason").[68] It is a wish on his part that also helps to explain why he continues to describe many of the tendencies that shape cinema, television, and the technologies of social networking in pessimistic, dystopic, moralistic terms:[69] because "contemporary technical mediation destroys the process of communication that once grounded orthographic writing." It does so by rendering the criteria of judgment by which the events to be mediated and retained are chosen and which makes memory precisely a question of politics—for Stiegler, that of a prejudged and predecided "calculable credit" (prejudged and predecided not least because on the networks, the "thinking who ... cannot think fast enough and must automate the process of anticipation").[70] In fact, the overall tenor of his message regarding digital media tends to be quite one-sided, even though he stresses at various points that he was interested in Web issues before the Web itself existed and that digital media technologies, as well as being part of the problem of the industrialization of consciousness, also have the potential to give our current control economy different inflections—and could even provide the "framework for an industrial model of change" that has moved beyond the consumer age by generating "new attentional forms that pursue in a different manner the process of psychic and collective individuation."[71] The impression conveyed nonetheless is that it is primarily the technologies and techniques of writing, the printed letter and the book (facilitating as they do the "deep attention" Stiegler rather too uncritically follows Katherine Hayles and Nicholas Carr in attributing to them, in marked contrast to the supposedly shallow "hyperattention" often associated with digital technologies) that, at the moment, really provide a means of reinflecting the subordination of individual agency and subjective thought to the formalized, standardized, reproducible, and controllable routines of the cultural and program industries.[72] "For me writing books is a technic of the self," he declares.[73]

In keeping with his view of the technologies of reproduction as Platonic *pharmaka*, neither simply poisons nor cures, Stiegler is quite prepared to acknowledge that "writing can be deployed as a sophistic or disciplinary individualization," as he puts it in a section on the power of writing in *Taking Care of Youth and the Generations*, and that writing "as a critical space is obviously and simultaneously duplicitous, pharmacological—and thus 'critical' in *that* sense."[74] Nevertheless, even when Stiegler does refer to the affirmative, productive, generative potential of cinema, multimedia, and digital television, he conceives such possibilities in terms that are very much derived from writing, the book, literature, and notions of literacy. "The real problem," he writes when bringing "The Discrete Image" to a close,

> is to rethink or think otherwise what Hollywood has up to this point done in the domain of the culture industry, to which cinema and television belong. ... Technology is giving us the chance to modify this relation, in a direction that would bring it closer to the relation of the literate person to literature: it is not possible to synthesize a book without having analyzed literally oneself. It is not possible to read without knowing how to write. And soon it will be possible to see an image analytically: "television" and "text" are not simply opposed.[75]

A great artist or philosopher for Stiegler is somebody "really specific, singular—somebody who is recognized as a singularity who has created a new type of circuit on which other people can come and continue the circuits."[76] It is a description that applies without doubt to Stiegler himself in many respects, thanks to his transformation of the contributions to the preindividual fund of Plato, Heidegger, Husserl, Derrida, Leroi-Gourhan, Simondon, and many others, along with his creation and inventive use of concepts such as organology, nootechniques, technogenesis, psychotechniques, and psychopower. Nevertheless, much of what he writes is concerned with the importance and value of paying attention and taking "care," together with the need to address the issue of knowledge and its relation to subjectivity afresh in the era of digital reproducibility. As a result, the question arises, just as Stiegler, in his account of how Western philosophy has excluded its origins with technics, sees Heidegger as having forgotten Epimetheus in *Technics and Time, 1*, is there something Stiegler has forgotten (but which, by the very emphasis he places on forgotten origins, on paying attention and on taking care, he can help us remember)? Has he forgotten to pay enough attention to the fact that the publishing of papercentric articles, monographs, and multivolume series of books submitted to learned journals and scholarly presses does not take place today outside and apart from the domain of the cultural industries (if it ever did), but is itself heavily implicated in the control and homogenization of our thought, memory, consciousness, and behavior through its media technologies? In short, is it possible that Stiegler has neglected to pay sufficient critical attention to the cultivation of his own self and conditions of his own individuation: specifically, the way his subjectivity, his way of being and doing as a philosopher and academic, is born out of a relation to technics and time?[77] I am thinking in particular of that aspect of our rapidly changing media environment that is associated with the print journal and book publishing industry, and the assemblages or meshworks of economic, social, legal, technological, and infrastructural links and connections that help to shape and formalize the conditions in which knowledge and research can and cannot be created, performed, organized, categorized, published, and circulated.

Admittedly, Stiegler draws attention to the "growing danger" represented by the privatization of the Web and the attentional forms it constitutes. He does so because the issue "is first and foremost political," due to the fact that the Web has become the new space of "the articulation between psychic individuation and collective individuation, and the site of fights to control the latter."[78] Yet that part of the publishing industry responsible for producing traditional print-on-paper academic journals and books is hardly free from privatization. Consider the increasing dominance in the English-speaking world of the market-led model of a small number of transnational corporations. Reed Elsevier, Springer, Wiley-Blackwell, and Taylor & Francis/Informa are far more concerned with productivity, efficiency, instrumentality, and the pursuit of maximum profit than with increasing circulation and making knowledge and research available to those who need it. (Indeed, according to one newspaper headline, they make Rupert Murdoch look like a socialist.)[79] This is evidenced by their already extremely high and still increasing journal subscription charges for those in science, technology, engineering, and mathematics (STEM) especially;[80] the "Big Deal," multi-year contract-bundling strategies that insist institutional libraries buy large numbers of publisher-generated packages of journals and prevent institutions from canceling subscriptions to even a single title; and the protection of copyright and licensing restrictions, not least through their support for measures such as SOPA (Stop Online Piracy Act) and PIPA (Protect IP Act) in the United States. Such policies led to an "Academic Spring" in 2012, whereby over 12,000 academics signed a public petition protesting the business practices of the largest of these megapublishers, Elsevier.[81] In contributing to the petition, academics pledged not to support Elsevier journals by publishing in them or undertaking editorial and peer-review work for them unless Elsevier withdrew its support for the Research Works Act, aimed at curbing government-mandated open access policies in the United States.[82] The Academic Spring was followed by a call to boycott both Taylor & Francis and Routledge if their parent company, Informa plc, does not bring down its journal subscriptions charges and pay the UK Exchequer the approximately £13 million (at the time of the call) lost to the treasury as a result of its 2009 decision to become a Jersey company domiciled in Zug, the canton with the lowest rate of taxation in Switzerland.[83] (Informa can thus be placed alongside Amazon, Apple, Facebook, eBay, Skype, and Google on the list of companies that have aggressively avoided paying the standard rate of corporation tax in the UK.)[84] With over "half of Informa's total annual operating profit ... derived from academic publishing: £85.8 million" in 2010 (£106.3 million in 2014), and its journals alone providing "gross profit margins of over 70 per cent," such a boycott would have consequences for some of the most highly respected titles in the critical theory and radical philosophy fields.[85] They include *Angelaki: Journal of the Theoretical Humanities; Cultural Studies; Continuum: Journal of Media and Cultural Studies; Communication and Critical/Cultural Studies; Culture, Theory, and Critique; Feminist Media Studies; Parallax; Rethinking Marxism*; and *Women: A Cultural Review*.[86]

The related dismantling of the kind of enclosed, disciplinary publishing organization designed more to serve charitable aims and the public good—scholarly associations, learned societies, university presses, and other nonprofit publishers—provides still further evidence of the dangers of privatization facing that part of the publishing industry responsible for producing traditional print-on-paper academic journals and books.[87] The already high and still increasing costs of journal subscriptions, combined with cuts to library budgets, subsidies, and other sources of funding, has "strangled libraries and led to fewer and fewer purchases of books/monographs."[88] This has produced a "monograph crisis," shorthand for the way the already uncertain sustainability of the print monograph is being placed at further risk by the ever-decreasing sales of such books.[89] The fall in demand for academic monographs has in turn resulted in presses producing smaller and shorter print runs. As a result, those volumes that are published are not distributed as widely as they may have been in the past, with many going out of print after only eighteen months.[90] Presses have also tended to favor publishing monographs from established academics who already have a strong readership, if not intellectual stars, rather than developing the next generation of scholars, whose sales are initially likely to be low yet who need to publish a research-led volume if they are to get a foot on the career ladder and acquire that all important first full-time position. Traditional print scholarly publishing therefore cannot be said to be explicitly dedicated to promoting the longevity, heritage, and intragenerational transmission from old to young—a process that for Stiegler forms an integral part of the production and selection of preindividual funds and one he describes as being "the essence of education."[91]

So hostile has the situation for publishing organizations designed to serve the long-term public good become that many of them are being forced to open up their walled gardens to the market and operate as if they were profit-maximizing businesses themselves. In fact (and as is the case today with regard to public infrastructures and services in general), a good number of them are being handed over to the corporations, in part or in whole.[92] They are thus finding themselves in the position of having to make decisions about what to publish (and consequently of having a major voice over who gets to have a career as an academic, researcher, theorist, or philosopher and who does not) more on the basis of the market and a given text's potential value as a commodity and less on the basis of its quality and value as a piece of peer-reviewed, properly referenced disciplinary scholarship and research. Some publishers are even moving much of their focus away from advanced level, full-length research monographs—especially those perceived as being radical, experimental, interdisciplinary, or avant-garde or deal with areas of thought regarded as particularly difficult, specialized, or obscure—to concentrate on textbooks, readers, introductions, reference works, and more fashionable, commercial, marketable titles. There has been a boom in the United States and UK, for example, in short academic/trade books focusing on particular films and TV programs, such as *Lost in Translation* and *Dr Who*, scholarly publishers thus tying themselves ever closer to the cultural industries and the system they form "with industry as such, of which the function consists in manufacturing consumption patterns by massifying life styles."[93]

When it comes to the threat of privatization and fights to control the space of articulation between psychic individuation and collective individuation, print and the Web cannot be simply contrasted in terms of an offline-online dialectic. Concepts, values, and habitual practices inherited from the era of writing, the book, and especially the industrialization of printing that took place from the middle of the eighteenth century onward—the indivisible and individualized proprietorial author, mass printing, uniform multiple-copy editions, fixity, the long-form argument, originality, authors' rights, copyright, and so on—are far from providing an unproblematic means of countering the businessification of the contemporary university. In fact, these historically inherited concepts, values, and practices also constitute some of the main ways in which knowledge, research, and thought are being commodified and corporatized by publishers of academic work—publishers whose business models now very much depend on turning even the publicly funded labor of radical philosophers such as Stiegler into marketable commodities.[94]

Nevertheless, as Dymitri Kleiner makes clear, authors and artists "continue to be flattered by their association with the myth of the creative genius, turning a blind eye to how it is used to justify their exploitation and expand the privilege of the property-owning elite." It is a state of denial and delegation of decision making (it could even be called cowardly in Stiegler's language) that has profound consequences for how we live, work, act, and think as theorists and philosophers:

> Copyright pits author against author in a war of competition for originality. Its effects are not just economic; copyright also naturalizes a certain process of knowledge production, de-legitimizes the notion of a common culture, and cripples social relations. Artists are not encouraged to share their thoughts, expressions and works, or to contribute to a common pool of creativity. Instead, they are compelled to jealously guard their "property" from others who they view as potential competitors, spies and thieves lying in wait to snatch and defile their original ideas.[95]

From this point of view (not to mention that of Anonymous, the *indignados*, and Occupy with which we began), many of the tendencies of which the current political economy of philosophy and theory is composed appear as yet another branch of the contemporary cultural industries: not just as some theorists and philosophers managing to "individuate themselves more intensely than others, and in doing so contribute more than others to the collective individuation"[96] but as some theorists and philosophers also acting as entrepreneurs and entrepreneurs of themselves, as Lazzarato puts it following Foucault, to market and promote their texts and make sure that the original ideas they contain (e.g., concerning new materialism, media archaeology, object-oriented philosophy, accelerationism, and so forth) are attributed to them as their (intellectual) property, thus both enclosing and branding these texts and ideas by association with a proper name.[97] If we do follow Stiegler in taking the idea of the control society seriously, we can see that they are engaged in "a war of competition for originality," implicitly and explicitly fighting with other critical thinkers over the "modulating principle of individual performance and merit" that "runs through each" (as measured by the amount and quality of publications, keynotes, and other indicators of reputation, impact, influence, and esteem), in order to gain advantage in the struggle for publishing opportunities, book contracts, jobs, promotion, grants, sabbaticals, support, resources, attention, recognition, fame.[98]

All of the above raises a number of questions regarding how Stiegler acts out what it is to be a radical philosopher for all he is frequently operating across the different publishing and academic systems of the

English- and French-speaking worlds. In this respect, monographs also appear as machines for "the production of ready-made ideas, for 'clichés,'" whose "criteria of selection are aspects of marketability." Monographs too are a means of standardizing and controlling thought, memory, and behavior (e.g., regarding authorship, originality, author's rights, copyright, intellectual property) "through the formatting and artificial manufacturing" of the desires of the individual theorist or philosopher, including those for preeminence, authority, and disciplinary power.[99] Such desires (or drives, since for Stiegler "a desire presupposes a singularity")[100] do much to explain the situation whereby the majority of even politically radical authors are willing to turn a blind eye and concede to the insistence of publishers that the rights to turn their text into a commodity that can be bought and sold (often for profit) be transferred to them: in exchange authors will have their work edited, copyedited, proofed, typeset, formatted, designed, published, distributed, marketed, promoted, and sold, and thus, they hope, read, recognized, and engaged with by others. In continuing to invest his time, care, and attention so heavily in the writing and publishing of conventional print codex books, can Stiegler be said to be exhibiting some of the very herd-like behavior, the generalized herdification, he condemns the cultural and program industries for producing in consumers in his essay, "How the Cultural Industry Destroys the Individual"? Is this not a variation on the "*liquidation of the exception*"? By being deprived of their individuality in this fashion, are even radical theorists and philosophers such as Stiegler, like the consumers of hyperindustrial capitalism, "lacking becoming, that is, lacking a future"?[101] In short, is there insufficient scope here too for the event, for singularity, for the welcoming of the new and opening of the undetermined to the improbable?

Wanted: Radically New Ways of Being Theorists and Philosophers

If Stiegler is right, then, and, with the Web and digital reproducibility, we have indeed embarked on a "radically new stage of the life of the mind whereby the whole question of knowledge is raised anew," this clearly has implications for our understanding of digital media technologies.[102] Just as important, it also has significant implications for our own ways of creating, performing, and circulating knowledge and ideas as critical theorists and philosophers. Not least, it suggests we need to be open to forms and techniques of analysis and critique that do not privilege writing and the associated acting out of the self as somehow separate from those technologies that provide it with a means of expression: paper, the book, film, photography, the Web, mobile media. Rather, it requires us to be open to what I would understand as more ethical and political forms of analysis and critique that welcome the new by helping to generate subjectivities that are different from how we currently live, work, act, and think. This includes ways of being theorists that depart from the self-disciplining neoliberal model of the entrepreneurial academic associated with corporate social media and social networks. However, it also includes ways of being that are different from the traditional, romantic, humanist, liberal model, with its enactment of clichéd, ready-made ideas of authorship, originality, the book, intellectual property, and copyright. In their own ways, both of these models are involved in the subordination of our agency and consciousness to the calculable, controllable, preprogrammed patterns and routines of the contemporary cultural industries. This leaves us with the following question: What forms might such different ways of creating, performing, and circulating theory and philosophy actually take?

One starting point for thinking about how we might address this question is provided by Lawrence Liang in his essay "The Man Who Mistook His Wife for a Book":

> To assert "This is my poem" within the social imaginary of intellectual property is to make a claim that sounds very much like "This is my pen," whereas in fact, it might be more accurate to think of its claim as the same as "This is my friend." And it is in this liminal space where poems look like pens that friendships get lost and property takes over.
>
> ... We can perhaps think of intellectual property rights as ... founded on very particularized ideas of property and personhood, but narrated as universal truths, that prevents us from seeing our own acts of reading, writing, creating, sharing, and borrowing in terms of the relational world that they occupy. Instead, we see them abstracted of their social relations.[103]

In order to challenge the European political, moral, juridical, and psychological individualism and sense of the "unified self" (285) that informs much of the Western metaphysical tradition, Liang proceeds to recount how "Indian culture does not draw a distinction between an agent who performs an action and the action that the agent performs." Instead, "an agent is constituted by the actions that he or she performs, or an agent is the actions performed and nothing more" (286).

So what forms might critical theory take when the focus is not only on what a theorist writes but also on the theory he or she acts out and performs? It is this question that is addressed in the following chapter. And since we are discussing how theorists act, it seems only proper and fair to explain the philosophy I am endeavoring to perform in some of the projects and actions I am involved with as an agent. As a way of embarking on the process of doing so, chapter 4 begins with an introductory overview of two projects in particular. The first is Open Humanities Press (OHP), a non-profit open access publisher I cofounded with Sigi Jöttkandt and David Ottina. The project I describe in most detail, however, is a series of experimental open access books I coedit with Clare Birchall and Joanna Zylinska as part of OHP: Living Books About Life.

4 THE POSTHUMAN

What Are the Digital Posthumanities?

How might we begin to produce not just new ways of thinking about the world, which is what theorists and philosophers have traditionally sought to do, but new ways of being theorists and philosophers too? I refer in particular to ways of being that do not remain unwittingly bound up with humanism and the humanities in the performance of their attempt to think through and beyond them. I take as my starting point for speculating on these questions here two projects with which I am engaged on a *practical* level: Open Humanities Press and the Living Books About Life series.

Part 1: Open Humanities Press

In 2013 the White House Office of Science and Technology Policy announced it was directing all federal agencies with a research budget of over $100 million to produce plans to make the results of their research accessible to the public within a year of publication. The White House's directive followed the 2012 decision of the European Commission to embed the open access principle into Horizon 2020, the European Union's Research and Innovation funding program for 2014 to 2020. Yet despite the fact it has clearly reached the mainstream, open access continues to be dogged by the perception that online publication is somehow less credible than print and that it lacks rigorous standards of quality control.[1] In the humanities, this view often leads open access journals, and book presses in particular, to be regarded as less trustworthy and desirable places to publish. They are seen as being professionally risky, especially for early career scholars. It is this perception of open access that Open Humanities Press (OHP), an international, nonprofit publishing collective specializing in critical theory, was set up to counter.[2]

Established in 2006 by Sigi Jöttkandt, David Ottina, and me, and working cooperatively with a wider network of scholars, librarians, technology specialists, and publishers, OHP began as a consortium of existing open access, online-only journals in philosophy, cultural studies, literary criticism, and political theory. These journals included *Cosmos and History*, *Culture Machine*, *Fast Capitalism*, *Fiberculture*, and *Vectors*. (Having started with seven, OHP has seventeen journals at the time of this writing.) While all these journals could be considered to be of good scholarly quality, many were at that time experiencing difficulty in generating a correspondingly high level of esteem because they were online journals rather than print and because most were then still relatively new. As Peter Suber observes, "new journals can be excellent from birth, but even the best cannot be prestigious from birth."[3] One of the drivers behind setting up OHP was a desire to bring these journals together under a single organizational umbrella in order to raise their profile and level of prestige in the eyes of academics and administrators by way of a meta-refereeing process. To this end, OHP put together an editorial board that includes, among others, Alain Badiou, Ortwin de Graef, Steven Greenblatt, Donna Haraway, N. Katherine Hayles, Bruno Latour, Alan Liu, Brian Massumi, Antonio Negri, Gayatri Spivak, and Ngũgĩ wa Thiong'o, and an editorial oversight group consisting of a rotating body of thirteen scholars drawn from the editorial board, that OHP utilizes to help it assess its titles according to a set of policies relating to publication standards, technical standards, and intellectual fit with its mission.

OHP officially launched in 2008. The plan at that time was to spend the first few years establishing a reputation for the press with its journals before proceeding to tackle the far more difficult problem of publishing book-length material open access. Things developed much faster than anticipated, however. As soon as OHP launched, it was contacted by a number of academics wanting to know when the press was going to publish books on an open access basis too. No doubt this was because books, and monographs especially, are extremely important to the humanities. Rightly or wrongly, they continue to be the gold standard by which the careers of academics in most parts of the humanities and many areas of the social sciences (HSS) too are judged.[4] And so in 2009 OHP established a monograph project that, in its start-up period, was run in collaboration with University of Michigan Library's Scholarly Publishing Office—what later became known as MPublishing.[5] The aim of this project was, and still is, to move forward both open access publishing in the humanities and the open access publishing of humanities monographs by experimenting with a model for the latter that does not rely on author- or funder-pays models. It thus does not risk disenfranchising independent scholars, those in less wealthy institutions, or those with alternative viewpoints that do not necessarily meet with institutional gatekeeper approval, be it at funding agency, university vice chancellor or provost, or research committee level—all individuals and bodies that would be deciding whether to pay the book processing charge (BPC) to the publisher in each particular case.

OHP initiated its monograph project with a number of high-profile book series. The most active of these now are New Metaphysics, edited by Bruno Latour and Graham Harman; Critical Climate Change, edited by Tom Cohen and Claire Colebrook; and Liquid Books, edited by Clare Birchall and me. By the end of 2011, OHP had begun publishing its first books.[6] Many titles have appeared since,[7] and three more series have also been added to the OHP roster: the Fiberculture Book Series, edited by Andrew Murphie; Immediations, edited by Erin Manning and Brian Massumi; and Technographies, edited by Steven Connor, David Trotter, and James Purdon.

It needs to be acknowledged that the vast majority of open access publications today, issuing from various publishing houses and other organizations, are still perfectly recognizable as journals, journal articles, and books in the conventional print-on-paper sense. The only difference—and it is an important one—is that they are made available online for free on an open access basis. For the most part, OHP is no exception to this rule. However, we are also interested in exploring the new forms that scholarly communication can take in the era of online publishing and open access. With this in mind, in 2013 we set up OHP Labs, a space for the OHP community to experiment with developing future models of theoretically informed critique. OHP Labs currently contains two projects: Feedback, a theory-driven humanities weblog publication edited by Henry Sussman and Jason Groves at Yale University and the Living Books About Life series.[8] I will come back to say more about some of the ideas that underpin OHP toward the end of this chapter. However, for now I turn to the second of these OHP Lab experiments: Living Books About Life.

Living Books About Life

Edited by Clare Birchall, Joanna Zylinska, and me and supported with an initial modest grant from Jisc, Living Books About Life was launched in November 2011. It is a series of open access books about life—with "life" understood both philosophically and biologically—that provide multiple points of connection, interrogation, and contestation between the humanities and the sciences. The twenty-five books that currently make up the series have been created by a globally distributed network of artists, theorists, and philosophers, including Mark Amerika, Claire Colebrook, Gabriela Méndez Cota, Alberto López Cuenca, Sarah Kember, Timothy Lenoir, Steven Shaviro, Oron Catts, and Ionat Zurr. The books repackage existing open access science-related research content from repositories such as ArXiv.org, PLoS (Public Library of Science), and PubMed Central by clustering it around selected topics, such as air, bioethics, biosemiotics, cognition, creative evolution, extinction, and human genomics, to form a series of coherent, single-themed volumes. By initially creating twenty-one "living books about life" in just seven months—which means the project is also an example of speed writing and editing to set alongside Tom Scheinfeldt and Dan Cohen's crowd-sourced book, *Hacking the Academy* (although theirs was of course only one volume, even if, to be fair, it was put together in just a week)—Living Books About Life offers one possible model for publishing books in the humanities in a relatively quick, easy, and low-cost manner.[9]

A number of factors drove this project, and especially its experimental side. Living Books About Life was for us a simple and inexpensive way of embracing some of the more imaginative forms of publishing that digital media make possible for the scholarly book-length argument, and all the more so when it is released from the majority of commercial marketing concerns and instead approached cooperatively by distributed communities of researchers, using content that is readily available online in shared open access spaces.[10] It is also the case that we have been advocating for academic research to be made available on an open access basis for some time, giving talks as well as writing books and journal articles on the subject. While activities of this kind continue to be extremely important to us, with the Living Books About Life project we wanted to do something other than just lobby for open access. We also wanted to engage, actively and practically, a number of humanities scholars—scholars who initially were not necessarily all that familiar with open access—in the various processes involved in publishing in this manner; that is, in living through these process, as it were. The idea was for us both to show those scholars how to publish their own work open access and to encourage them to make greater use of the work of others that is already available in open access journals and in central, subject-based, and institutional repositories, not least by drawing on it in their research and teaching (by referencing it, citing it, including it in bibliographies and reading lists, and so on). Just as important perhaps, we wanted to increase the awareness of those scholars and their readers of the consequences of *not* making research available open access.

The process of creating such a "living book" for the series was as follows. Having decided on their science-related overall topic concerning life (astrobiology, say, or symbiosis), the editors—all humanities scholars who had prior experience of working on science topics—were asked to search various open access science journals and repositories with a view to selecting at least ten items they wished to include in their book. This number was chosen in order to ensure that a large part of each living book contained content that had already been peer-reviewed—although, once they embarked on the process, most editors exceeded the stipulated minimum. The editors were also invited to supplement their selected open access research content with additional material: peer-reviewed book chapters, as well as extracts from books, pages, snippets, and quotations but also, so as to make full use of the wiki platform on which the project was hosted, images, infographics, podcasts, and videos. With this we were looking to test some of the physical and conceptual limitations of the traditional codex volume. Thus, a number of books in the series themselves included whole books, for instance. One editor even went so far as to produce two tables of contents: a standard printable table and a second table in the form of an interactive map, containing article summaries, that was geolocated with surveillance cameras.[11]

The editors were then asked to write an opening essay, linking the content of their volume together and providing a point of connection and translation, as well as possible interrogation and contestation (thus avoiding the trap of "scientism"), between the not unproblematic two cultures: the humanities and the sciences. Although a significant proportion of each book was made up of open access material that had already been peer-reviewed, the project team still arranged for each living book to be peer-reviewed as an entity, in line with the standard practice for edited volumes in the humanities. The project team also copyedited, proofed, and formatted the books; provided front and back *cover pages*; assigned ISBN numbers; and, in response to a request from some of those involved, created *frozen* PDF versions of how each book initially stood before it was made available for reediting by other users.

The books were published on an open source wiki and blog, using open source software: MediaWiki by Wikimedia and WordPress. Since figures indicate WordPress is used by 21.5 percent of all websites, most scholars are familiar with it, especially with its simple WYSIWYG interface and its cut-and-paste techniques known from word processing packages.[12] MediaWiki, however, is a free open source software wiki package written in PHP, originally for use on Wikipedia. It has since been adopted and used by several more projects of the nonprofit Wikimedia Foundation and by many other wikis, including MediaWiki itself and, of course, WikiBooks.[13] Simple instructions were provided to all editors to help them with creating their books online. However, no technical knowledge was required: if the editors were able to use a word processor of any kind (Microsoft Word or an editor from the Open Office suite), they were equally able to edit their living book.

One of the most significant and revealing aspects of the Living Books About Life series for us was the impact it had on the researchers who took part in it. Embarking on the project, we quickly discovered that there was a relatively low level of knowledge among our humanities-based editors about what exactly was entailed in publishing open access and about the restrictions that were often placed on academic publications by publishers' licensing and copyright agreements. For example, it took a number of our initial editors quite some time to realize that while they might be able to openly access a given article online from their computer at work because, unbeknown to them, their institution had subscribed to a particular journal or content provider such as JSTOR or Project Muse, others whose institutions had not taken out such a subscription or who were not affiliated with an institution might not be able to access the same article online at all. We also found confusion concerning the difference between open access articles made available under Creative Commons licenses that did allow others to copy, reuse, distribute, transmit, and display them publicly and to make and distribute derivative works, and articles using proprietary licenses that, although they may indeed have been deposited in an open access repository such as PubMed, were nevertheless not available for reuse and could consequently be made accessible only via a direct link. (Many of the scholars involved in the project were surprised at how much of the research they were keen to include in their living books was not available on an open access basis and that still less was available under a license, such as CC-BY, that would allow others to copy and reuse it.)[14] For purposes of clarification with regard to copyright and intellectual property (IP), each of our editors was asked to supply licensing information for all the articles and other items included in their books. This took the form of an attributions page, providing—for every entry—an exact reference, a URL address and information about any license, copyright, or permission required and obtained. (Such information was normally available on each original article included. If not, the editor in question would have taken the appropriate steps to obtain permission from the copyright holder—for example, the author or publisher of the journal in which the article had appeared.)

The educational aspect of the Living Books About Life project with regard to copyright and intellectual property issues was certainly another of the motivating forces behind it. We wanted to experiment with publishing books not just on a gratis (free) open access basis, but on a libre (reuse) basis too. The Budapest-Bethesda-Berlin definition of *open access*, which is one of the major agreements underlying the open access movement, lists the ability to reuse material (and not just access it) as one of its essential features.[15] Yet research shows that "of the books presently available open access, only a minority have a license where *price and permission* barriers to research are removed."[16] Consequently (and as chapter 1 illustrates with the example of re.press and Graham Harman's *Prince of Networks*), although "the right to re-use and re-appropriate a scholarly work is acknowledged and recommended, in both theory and practice a difference between 'author-side openness' and 'reader-side openness' tends to be upheld."[17] By contrast, the books in the Living Books About Life series do not provide readers merely with the ability to comment on, respond to, interrupt, debate with, and challenge the text, the author, and other readers; these "living books about life" are themselves living, in the sense that they are open on a read-write-rewrite basis for users to help compose, edit, annotate, translate, distribute, and remix them, should they so wish, as well as create derivatives. What this means is that together with repackaging (in his book on uncreative writing in the digital age, Kenneth Goldsmith refers to processes of this kind as *manipulating* and *managing*)[18] the available open access material on life into a series of books, the project is also engaged in rethinking the book itself as a living, collaborative, processual endeavor in the era of open access, open data, open science, and open education.[19] Anyone with access to the Internet can get involved in creating books for the series potentially or in adapting versions of existing books for their own use. The books can thus become a teaching and learning aid, an alternative form of online course pack, course reader, or Open Educational Resource (OER), one where the content and form of the book can be negotiated, updated, and altered by learners themselves, under the guidance of a tutor (or not, as the case may be).

With the Living Books About Life series, we were therefore more interested in exploring the processes, mechanisms, and concepts involved in publishing than in making use of and testing particular tools, platforms, or products. This is why we chose the straightforward and simple-to-use technology of a wiki. Even in its very relational connectedness, a lot of Web 2.0

culture, such as that associated with Facebook, Twitter, and many blogging tools, is quite humanist, individualistic, and neoliberal in the way it is structured and performed.[20] Wikis, however, tend to be the product of groups or networks that it is often difficult to assign a fixed or unified identity to but are nevertheless working cooperatively, collaboratively, and at times anonymously. These groups and networks manage to build hard-to-recreate resources for free, with a view to generating texts that are both created by the community and retained by it.[21]

Variations on the Living Books About Life model have since been adopted by a number of other open access publishers, including Punctum Books in New York and Springer's book, *Opening Science*.[22] However, the course reader that Joanna Zylinska (also one of the editors of the Living Books series) developed in 2010 for Open Humanities Press's longer-running sister project, Liquid Books, again edited by Clare Birchall and me, is one of the more interesting examples of this philosophy in action. This reader was used for a ten-week graduate theory course, Technology and Cultural Form: Debates, Models, Dialogues, taught in a workshop format at Goldsmiths, University of London. The course discussed the relationship between various media and technological forms, their social uses, and the culture in which they operate. In this context, the "liquid reader" provided a practical case study of a media form that students were able to both think about and actively construct. A basic skeletal course reader was first created online using the wiki platform. It included the key course content and was subsequently opened to customization by students. Throughout the course, students were involved in adding and editing the reader's content. They were also encouraged to experiment with the idea of the reader—or, more broadly, the book—through activities such as collaboratively writing a wiki-style introductory, self-reflexive essay titled "Future Books: A Wikipedia Model?" and putting together an online gallery of their photographic work on the course topic as part of the reader. The idea was to provide an open access, OER study tool that facilitated the sharing of knowledge and pedagogic practice. The resulting course reader continues to be freely available to both Goldsmiths students and to students, tutors, and the community of general users internationally.[23]

Postindividualistic

Given that many of the ideas on which the humanities are based—the unified, sovereign, proprietorial subject; the individualized author; originality; the finished object; immutability; copyright; and so on—are commonly held as a means of sustaining authority and creating trust between the author and reader, the question arises as to why we would put this authority and trust at risk by making the books in our series available on an open read-write-rewrite basis. After all, many would argue that academic books, and monographs especially, with their ability to allow authors to develop complex arguments and analyses over an extended span, are, in the words of Nigel Vincent, British Academy vice president for research and higher education policy, "an intrinsically important mode of academic production." Monographs are certainly too important to be "sacrificed on the altar of open access." In fact, Vincent goes so far as to insist that "adoption of the untrammelled CC-BY licence," while suitable for areas of science that need to be able to search, recover, and mine large sets of data, "is not appropriate for monographs and book chapters" at all.[24]

One way to explain the philosophy behind the Living Books About Life project and its spinoffs is in terms of the concept of the digital posthumanities (which should really perhaps be called *postdigital* post*humanities*, if it were not such a clunky term). Our interest in the posthumanities goes at least as far back as 2006 and "Cultural Studies and the Posthumanities," an essay that Birchall and I commissioned Neil Badmington to write for a book called *New Cultural Studies: Adventures in Theory*.[25] Badmington's piece allows us to see that this idea of the digital posthumanities builds on research conducted not just within the digital humanities but under the sign of the posthuman too; but it also marks a departure from much of this research. To articulate exactly how and why, I shift the focus of attention of this chapter to a recent book by Rosi Braidotti, *The Posthuman*, which provides what is undoubtedly one of the most interesting and thought-provoking examples of posthuman studies today. Like *New Cultural Studies*, the Living Books About Life series, and Open Humanities Press, it is very much concerned with "the status and value of theory itself."[26] Indeed, *The Posthuman* is located precisely at the intersection between the posthuman and the humanities, trying, testing, and troubling the latter's limits, yet it also draws attention to some of the other, nonunitary and "postindividualistic" ways of being and living as theorists that are possible today.

Part 2: Posthumanism

Posthumanism is concerned with the displacement of the unified, self-reflexive, and rational humanist subject from its central place in the world as a result of the erosion of the human's "natural" boundaries with the animal, technology, and the environment. Braidotti identifies one of the main questions she wants to address in her book as focusing on how the practice of the humanities today is affected by the posthuman—in particular, What is the function of theory in posthuman times? (3). Braidotti answers this question by suggesting that the task of critical theorists is to present suitable "representations of our situated historical location" (4). This cartographic endeavor—a cartography for Braidotti being "a theoretically based and politically informed reading of the present" (164)—is bound up with the notion that the line dividing ideas of the natural and the cultural has been displaced and to a large extent rendered indistinct by the effects of scientific and technological advances, including those associated with robotics, prosthetics, artificial intelligence, and genetic engineering. *The Posthuman* therefore takes as its starting point the belief that social theory must respond to the change in concepts, methods, and political practices that is being brought about by an associated paradigm shift. It is a shift from a social constructivist approach, which proposes a fundamental difference between what is considered to be given (nature) and what is constructed (culture), to a monistic philosophy, derived from the work of Baruch Spinoza, Henri Bergson, Gilles Deleuze, and Félix Guattari. It is a philosophy "which rejects dualism, especially the opposition nature-culture and stresses instead the self-organizing (or auto-poietic) force" that is present in all living matter (3).

Braidotti's golden rules and methodological guidelines for posthuman critical theory include, first, what she calls cartographic accuracy, which aims at epistemic and ethical accountability by revealing and mapping the locations of power that structure our subject-position. This form of accountability is important because it allows theorists to account for their own location "in terms of both space (geo-political or ecological dimension) and time (historical and genealogical dimension)" (164). Stressing the situated aspect of critical theory in this manner helps theorists to understand how they are implicated in the very social arrangements, frames, and discourses they are endeavoring to analyze and oppose. It also draws their attention to the fact that all knowledge claims are partial or limited in nature, something that is crucial as far as any critical engagement with liberal individualism and universalism is concerned.

Braidotti's second methodological guideline is that critiques of such power locations should work alongside the search for creative "alternative figurations or *conceptual personae* for these locations," which "express representations of the subject as a dynamic, non-unitary identity" and which dramatize "the processes of becoming" (164). The examples Braidotti offers—although, in keeping with her emphasis on the situated nature of critical theory, she highlights these articulate "complex singularities, not universal claims"—include "the feminist/the womanist/the queer/the cyborg/the diasporic, native, nomadic subjects" (164).

The third of Braidotti's rules for posthuman critical theory that merits particular attention is the adoption of zigzagging, nonlinear thinking in response to what she sees as the

> complexity of contemporary science and the fact that the global economy does not function in a linear manner, but is rather web-like, scattered and poly-centered. The heteroglossia of data we are confronted with demands complex topologies of knowledge for a subject structured by multi-directional relationality. We consequently need to adopt non-linearity to develop cartographies of power that account for the paradoxes of the posthuman era. (164–165)

Braidotti proceeds to identify a number of criteria for a new posthuman ethics that would be capable of generating, creatively and affirmatively, conditions for revitalized political and ethical agency in the Anthropocene era. These criteria include "non-profit; emphasis on the collective; acceptance of relationality and viral contaminations; concerted efforts at experimenting with and actualizing potential or virtual options; and a new link between theory and practice, including a central role for creativity" (191). At the same time, she argues that the "contemporary university needs to redefine its posthuman planetary mission in terms of a renewed relationship to the global city where it is situated." The key terms for Braidotti in this process are (a rather vague and undefined) "open source, open governance, open data and open science, granting free access by the public to all scientific and administrative data" (180). (The last is also one of the political responses to the increasing instrumentality and functionality of society postulated by Jean-François Lyotard at the end of *The Postmodern Condition*.)[27]

What makes *The Posthuman* such an important intervention in the current politico-intellectual conjuncture, and the reason I am interested in looking at it here in the context of a discussion of the future of theory and the humanities in the twenty-first century, is the way Braidotti is clearly opening the door for a radical mutation of many of the concepts and practices on which theory and the humanities are currently based: the subject as a static, stable, unitary entity; the indivisible and individualized author; the linear argument and text; originality; the signature; the finished object; fixity; copyright—and, indeed, the very concept of the human. She thus seems to be calling for a profound transformation in our ways of acting, thinking, writing, and speaking as theorists and philosophers. And to the extent Braidotti is calling for such a transformation, I agree with nearly everything she says in *The Posthuman*. I absolutely concur with her emphasis on the situated structure of critical theory; on the search for alternative, affirmative figures or personae for these locations that dramatize the processes of becoming; on "experimenting with and actualizing potential or virtual options" (191)—for a dynamic, nonunitary, "expanded, relational self" especially (61)—and the role both critique and creativity have to play in such processes; on nonprofit, nonlinearity, collectivity, open source, and open science; on creating a new polity or community; and on being "active in the here and now of a continuous present that calls for both resistance and the counter-actualization of alternatives" (192).

And yet at the same time as she is opening the door for such a transformation in our ways of being theorists and philosophers in *The Posthuman*, Braidotti can on a number of occasions also be seen to keep the security chain on, if not actually close the door again. "We do need to embrace nonprofit as a key value in contemporary knowledge production," she emphasizes, linking this "gratuitousness" to the "construction of social horizons of hope" (185). Yet this aspect of Braidotti's new posthuman ethics, and the institutional practice she regards as being "best suited to posthuman critical theory and to the twenty-first century Humanities," is rather undercut by the fact that *The Posthuman* has not actually been published on a nonprofit basis at all (173). Instead, it has been brought out by Polity Press, which is an independent but nevertheless for-profit press, distributing *The Posthuman* through John Wiley & Sons, one of the "big four," profit-maximizing, scholarly publishing megacorporations, along with Reed Elsevier, Springer, and Taylor & Francis/Informa.[28] Nor, by doing so, has Braidotti made *The Posthuman* available to others to be legally shared and reused on an open basis, be it in terms of open source, open science, or open access. In fact, for all that the posthuman is supposed to introduce "a qualitative shift in our thinking," it is in many respects hard to tell the difference between Braidotti's book and most of the other critical theory monographs that have been produced in the last thirty years or so (2). *The Posthuman* certainly adheres closely to the classical (in the spirit of chapter 3's analysis of Stiegler, we might even call it *clichéd*) definition of a monograph as "a printed specialist book-length study ... written by a single academic author."[29] As such, it seems somewhat at odds with the emphasis Braidotti places on the need for posthuman theory to respond to the complexity of contemporary science. After all, as even Nigel Vincent, the British Academy vice president for research and higher education policy, makes clear in his discussion of the challenge he sees open access to be presenting to the academic monograph, the close relationship in the humanities "between the individual(s), the research and the writing is at the opposite pole from what goes on in some areas of the natural sciences, where in the extreme case there may be hundreds of names of 'authors' attached to the paper."[30] (Of course we might also ask, What is the imperative to respond to the complexity of science? Does science have the answers the humanities lack? How does responding to science avoid merely becoming a form of scientism?)

There is also little evidence of what Braidotti calls defamiliarization, the methodological "process by which the knowing subject disengages itself from the dominant normative vision of the self he or she had become accustomed to, to evolve toward a posthuman frame of reference," being concretely actualized in practice here (167). The posthuman nomadic subject may be "multifaceted and relational" (188) and have its basis in "transversal interconnections across the classical axes of differentiation" (96). But where are the examples in *The Posthuman* of Braidotti exploring some of the ways we can actually "experiment together with alternative forms of posthuman subjectivity" when it comes to how she herself acts according to the author-function (187)?

In fact, far from her methodology not being "about the authority of a proper noun, a signature, a canon, a tradition, or the prestige of an academic discipline," as she puts it at one point, *The Posthuman* very much functions to help certain established forms of authority and ways of being to persist (165). These include the forms of authority associated with the construct known as Braidotti-the-internationally-renowned-and-respected-theorist—a construct perhaps as different from Braidotti as, according to Bruno Latour, Pasteur-the-great-researcher was from the scientist composed of those elements that, through the links between them, formed the "Pasteur network," many of them objects and nonhumans such as notebooks, laboratories, microbes, and vaccinated cows.[31] *The Posthuman* also sustains the not unrelated sense of Braidotti as an identifiable, self-contained, rational individual human, whose subjectivity is static and stable enough for her to be able to sign her name on a contract giving her the legal right to assert her identity as the "Author of the Work … in accordance with the UK Copyright, Designs and Patents Act 1988," and to claim this original, fixed, and final version of the text as her isolable intellectual property—not least via an all-rights-reserved copyright notice.

When it comes to how Braidotti enacts her role and identity, her way of working, thinking, and writing as a posthuman critical theorist, then, we can see that for all the emphasis she places in her methodological guidelines and criteria for a posthuman ethics not just on nonprofit but also on open source, open science, collectivity, and the counteractualization of affirmative, postidentitarian alternatives to dominant representations of the self, we are still very much with the post-Enlightenment model of the individual human "subject as citizen, rights-holder, property-holder."[32] What Bradotti says of the humanities in general seems to apply to her as well in many respects: both appear "to be unable to resist the fatal attraction of the gravitational pull back to Humanism" (153).

Derrida versus Deleuze?

I am aware that some readers may be scratching their heads at this point over the seeming contradiction evident in the fact that the argument I am presenting here is being made in yet another conventional monograph, signed with a singular author's name, and published with a brand-name press. I will address this issue in relation to Braidotti (and Cary Wolfe) shortly by proposing an addition to her conceptual personae of the feminist, the cyborg, oncomouse, and Dolly the sheep. This will be an alternative figuration somewhat akin to that of the pirate. Its task will be to work alongside my critique of those locations of power that structure our subject-position in terms of the proprietorial subject, originality, copyright, and so on (a critique designed precisely to help us understand how our theory is implicated in the very social arrangements we are trying to oppose).[33] For now I just want to comment further on my reasons for focusing on Braidotti (a theorist who has been extremely important to me over the years). I hope readers will see the interpretation and thinking through of her position I am providing here (as well as my readings of Hardt and Negri, Harman and Latour in chapter 1), as more than a series of cheap shots. The reason I am raising these questions with regard to her approach is that if the "key for everything" for Braidotti "lies in the methodology," then it is important we pay close attention to her own methodology in *The Posthuman* (163)—and all the more so given that she considers a "concrete, actualized praxis" to be "the best way to deal with the virtual possibilities that are opening up under our very eyes, as a result of our collectively sustained social and scientific advances" (196).

To be fair, an unwitting reinstatement of humanism is something Braidotti shares not only with many of those in the humanities, but also with the majority of those associated with the posthuman too. Ivan Callus and Stefan Herbrechter, for example, follow Donna Haraway's "Cyborg Manifesto" in arguing that "the erosion of one human/non-human boundary inevitably leads to breakdowns in other boundaries."[34] What they dub "critical post-*humanism*" is thus interested in the "as read" aspect of the following question: "How can one read in a manner that does not take 'as read' the humanity from which one reads?" (96). In other words, critical posthumanism is a "humanism intent on working through its own represseds" (103). Accordingly, one might argue that what we are trying to develop with the Living Books About Life series is a critical posthumanities to the extent that we are interested in the "as read" in the slightly different question: "How can one read in a manner that does not take 'as read' the established configurations of the humanities from which one reads (their forms and methods and so on)?" For my coeditors and me, this critical look needs to include challenging the emphasis in both the humanities and posthumanism on reading and writing as the "natural" or normative practices through which such questions can be raised and addressed. Indeed, reading and writing could be said to represent part of critical posthumanism's own repressed. For while the human's boundaries with the animal, technology, and the environment—or any other nonanthropomorphic *others*, for that matter, including insects, plants, seeds, bacteria, the planet, and the cosmos—may all come tumbling down in posthumanist discourses, one dividing line tends to remain intact in arguments that enact such boundary crossing: that between the humanities and (their becoming) posthumanities. Even the most critical of posthuman theorists—those who are not simply viewing the future of the human in either utopian (as in the case of the extropians) or dystopian (e.g., Francis Fukuyama) terms—seem intent on studying the animal, technology, and the environment more with a view to undermining anthropocentricism and humanist essentialism than creatively exploring and experimenting with the core foundational concepts, values, forms, and methods of the humanities.[35]

Nor does the series launched by Cary Wolfe in 2007 explicitly on the subject of the posthumanities constitute an exception. There is much to admire about this series intellectually, including as it does titles by, among others, Roberto Esposito and Donna Haraway. In fact, its vision of posthumanist theorists moving beyond the "standard parameters and practices" of the humanities to think differently about themselves and what they do and pursue alternative schemes of thought, practice, knowledge, and self-representation, is an extremely exciting one.[36] I agree with Wolfe when, in his own contribution to the Posthumanities series, he argues that "one can engage in a *humanist* or a *posthumanist* practice of a discipline" and that a discipline has to be aware of its "*own* modes of disciplinary practice, its own forms."[37] This is why, for Wolfe, we need theory: because of its ability to help raise our awareness of such disciplinary practices and forms. Still, how does this pronouncement square with the fact that what this series has produced in the main is yet more specialist, linearly structured, sequentially paginated, book-length studies, published by internationally renowned individual academic authors as original fixed and final monographs, in uniform multiple-copy editions, on an all-rights-reserved basis, copyright of the Regents of the University of Minnesota? Is humanism not "actually being reinstated uncritically" here too?[38]

This tension between the humanities and the posthuman is something Braidotti herself does acknowledge a number of times. (Actually, I could have developed a similar, though not the same argument with regard to many of those associated with nonhumanist theories and philosophies. One of the reasons I find Braidotti's ideas in this book in particular so thought provoking and productive is that they are located precisely at this intersection between the anti-, non-, or posthuman and the humanities, pushing at the latter's limits far more than most.) With its emphasis on the subject as dynamic, nonunitary, and relational and on nonprofit, the collective, open source, and the processes of becoming, *The Posthuman* points us in the direction of a profound mutation of many of the forms and methods that lie at the heart of the humanities. But if the question is about what the humanities can become in the age of the posthuman, "after the decline of the primacy of 'Man' and of *anthropos*," the answer for Braidotti is most definitely not post*humanities* (173). And this is the case regardless of the fact she considers neither humanism nor antihumanism to be ultimately up to the job—posthumanism for her being "the historical moment that marks the end of the opposition between Humanism and antihumanism and traces a different discursive framework, looking more affirmatively towards new alternatives" (37). Braidotti is concerned with how the humanities can be inspired by "experiments in posthuman thought and new post-anthropocentric research," not with their becoming posthumanities (163). Rather, "Posthuman times call for posthuman Humanities studies," she declares (157). Or, to put it another way, Braidotti is interested in the posthuman, but from a more humanities than posthumanities perspective. Consider the way she writes that "the Humanities need to mutate and become posthuman, or to accept suffering's [sic] increasing irrelevance" (147). Markedly, she does not suggest that the humanities need to mutate and become posthumanities. Witness, too, the examples she gives of posthuman humanities studies—environmental humanities, digital humanities, cognitive or neural humanities—all foreclosed as belonging to the humanities. Braidotti thus pushes her book as close to the extremes of the humanities as she can without their actually becoming posthumanities. Once again we are dragged back toward (and by) humanism and the humanities.

Nor does Braidotti consider the contradiction between the humanities and the antihumanism inherent to posthuman critical theory to be a fundamental problem:

> The best example of the intrinsic contradictions generated by the anti-humanist stance is emancipation and progressive politics in general, which I consider to be one of the most valuable aspects of the humanistic tradition and its most enduring legacy. Across the political spectrum, Humanism has supported on the liberal side individualism, autonomy, responsibility and self-determination. On the more radical front, it has promoted solidarity, community-bonding, social justice and principles of equality. ... These principles are so deeply entrenched in our habits of thought that it is difficult to leave them behind altogether.
>
> And why should we? Anti-humanism criticizes the implicit assumptions about the human subject that are upheld by the humanist image of Man, but this does not amount to a complete rejection. (29–30)

In fact, as far as Braidotti is concerned, "one touches humanism at one's own risk and peril" (29). That is all very well, but it does beg the question: How does this continued support for humanism and the values and practices of the (postanthropocentric and posthuman) humanities relate to the importance she attaches to affirmative alternatives to dominant visions of the subject and self, to nonprofit, collectivity, open source, and so forth? If we accept that we live in posthuman times and *do* want to act according to the rules, guidelines, and criteria she sets out for posthuman critical theory and posthuman ethics, does this not require us to move beyond the standard parameters and practices of the humanities, as Wolfe's "Posthumanities" suggests?

Part 3: Zombie Materialism

The first reference Braidotti makes in *The Posthuman* is to this short (and now hard to find in its full version) text by Wolfe in which he argues that instead of "reproducing established forms and methods of disciplinary knowledge," posthumanists need to "rethink what they do—theoretically, methodologically, and ethically."[39] Braidotti mentions this proposition, however, not to explore the possibility of becoming posthumanities, but simply to draw on his description of what is meant by the human after the Enlightenment: "The Cartesian subject of the cogito, the Kantian 'community of reasonable beings,' or, in more sociological terms, the subject as citizen, rights-holder, property-holder, and so on."[40] Braidotti does not refer to Wolfe's "Posthumanities" again in her book. The only other time she mentions Wolfe[41] is in a discussion of the relation of the posthuman to the humanities that immediately follows the passage quoted above about the intrinsic contradictions generated by the antihumanist stance:

> The difficulties inherent in trying to overcome Humanism as an intellectual tradition, a normative frame and institutionalized practice, lie at the core of the deconstructive approach to the posthuman. Derrida opened this discussion by pointing out the violence implicit in the assignation of meaning. His followers pressed the case further: "the assertion that Humanism can be decisively left behind ironically subscribes to a basic humanist assumption with regard to volition and agency, as if the 'end' of Humanism might be subjected to human control, as if we bear the capacity to erase the traces of Humanism from either the present or an imagined future" (Peterson, 2011: 128). The emphasis falls therefore on the difficulty of erasing the trace of the epistemic violence by which a non-humanist position might be carved out of the institutions of Humanism. ... In this deconstructive tradition, Cary Wolfe is especially interesting, as he attempts to strike a new position that combines sensitivity to epistemic and word-historical violence with a distinctly trans-humanist faith in the potential of the post-human condition as conducive to human enhancement. (30)

Braidotti adopts this line of argument as further support for her decision to argue for the development of a posthuman humanities studies, rather than posthumanities, by way of moving beyond the contradictions and tensions between humanism and antihumanism.

It is interesting, then, that one place where the issue of the responsible decision and the violence implicit in the assignation of meaning has been raised in relation to *The Posthuman* is precisely with regard to Braidotti's reductionist and rather negative attitude toward philosophical theories associated with so-called poststructuralism and deconstruction.[42] (And this is in spite of what she says about wanting to avoid, or indeed transcend, negativity and to support a monistic philosophy that "rejects dualism" (3) in order to "overcome dialectical oppositions" and engender "non-dialectical understandings of materialism" (56).) Braidotti's complaint about critical thought "after the great explosion of theoretical creativity of the 1970s and 1980s" is that it was as if "we had entered a zombified landscape of repetition without difference" (5). I understand why she might say this—even if *zombified* does seem a rather negatively connoted word to use. Without doubt poststructuralist critical theory did in certain hands become another orthodoxy. (The usual move is to castigate literature departments in the United States as the chief offenders.) Still, if we are to make statements about the zombified landscape of theory, it is best to avoid succumbing to similar zombie repetitions ourselves, as much as can be reasonably expected given it is impossible to achieve perfect self-consciousness. Ironically, this is not something Braidotti is especially successful in doing in *The Posthuman,* as her comments about the "limitations" of deconstruction's "linguistic frame of reference" being the reason she prefers to take a more "materialist route" when dealing with the posthuman bear witness (30).

Oversimplified position statements of this kind are not confined to Braidotti's book, of course. In fact, if there is not one already, someone should establish a website or blog to record them all. They could do worse than begin with examples of the repetitive rhetoric that is often used to divide the history of critical theory into movements, moments, trends, or turns: the cultural turn, linguistic turn, affective turn, visual turn, computational turn, materialist turn, majoritarian turn, nonhuman turn, geological turn, and so on.[43] From there they could proceed to gather the associated attempts to replace one mode, orientation, or attitude of thought with another—textualism with realism and materialism; negative and oppositional external critique with constructive, creative, immanent affirmation;[44] representational with nonrepresentational theory; and the emphasis on lack in Lacanian psychoanalysis with the "desiring theory" of much "Deleuzianism"[45]—by declaring we "no longer" live in one era and now belong to another, be it that represented by the shift from hegemony to posthegemony, social constructivism to monism, or the "speculative turn" away from the previous "deconstructionist era" and the subsequent "period dominated by Deleuze" toward nonhuman ontologies and the Anthropocene.[46]

Yet I cannot help wondering whether Braidotti's repetition of certain reductive refrains regarding theory—despite the respect she professes for it and for her post-1968 teachers, who included Michel Foucault, Luce Irigaray, and Gilles Deleuze—is connected to the (non)decisions she makes over nonprofit, open source, and collective ways of acting, working, and thinking as a theorist, and about not pushing further toward becoming posthumanities. Consider this claim that Braidotti makes:

> The posthuman subject is not … post-structuralist, because it does not function within the linguistic turn or other forms of deconstruction. Not being framed by the ineluctable powers of signification, it is consequently not condemned to seek adequate representation of its existence within a system that is constitutionally incapable of granting due recognition. …
>
> The posthuman nomadic subject is materialist and vitalist.[47] (188)

What is being given yet another outing here, as Herbrechter quite rightly points out, is the by now all-too-familiar antagonism over the material and language, "affirmation and negativity, action and decision," between those approaches inspired by Gilles Deleuze and those more influenced by Jacques Derrida. Given the emphasis placed in *The Posthuman* on being both critical and creative, the issue is "where and at what level the 'critical' would 'bite"' (or indeed *cut*, as we shall see shortly). For those steeped in a rigorous engagement with the philosophy of Derrida—with whose name deconstruction is most closely associated—"this would at least also have to occur at the level of language (or discourse)."[48] This would in turn render problematic Braidotti's attempt to distance her theory of the posthuman subject from modes of critical thought concerned with representation, signification, and the linguistic:

> Not only does Braidotti here somewhat betray her own intellectual "cartography" but she is also arguably ridding the future humanities of their most important methodology on which, precisely, the critical potential of posthumanism will depend: namely making sure everyone remembers that the argument about the posthuman is fought precisely at the level of representation, symbolic meaning and thus (amongst other "media") in language.[49]

New Materialism: I Know *Zombie*'s a Bit Strong But ...

Many of the prejudices Braidotti displays in *The Posthuman* regarding theory have been accepted almost as a form of common sense in much of the humanities and social sciences for some time now. Thanks to a complacent adherence to this new orthodoxy, poststructuralism and deconstruction are regularly positioned by stands of critical thought associated with "new materialism" as being exactly the kind of transcendent, language, writing, and text-oriented theories we need to move on from in order to concentrate on those aspects of material reality our culture is increasingly regarded as being actually about (e.g., infrastructure, hardware, software, code, platforms, dashboards, interfaces and of course their physical supports and material substrates: cables, wires, chips, circuits, disks, drives, airwaves, electrical charges, optical rays). Dennis Bruining relates such new materialist discourses to the way in which, in spite of both the poststructuralist critique of foundations and their own awareness of the untenability of ideas of this kind (of biology as destiny, for example, in the case of theories of life, genetics, and the body), "there still lingers the notion of, and a longing for, a present underlying foundation and/or truth in some political and theoretical movements and writings." It is a longing for truth or foundation that Bruining connects to the contemporary turn to science in the humanities. But as Clare Birchall and I argue in our contribution to *New Cultural Studies*, attachments of this nature can also be linked to what

Wendy Brown calls antipolitical moralism. As we write there, this is a term Brown uses

> to refer to a certain "resistance" to thinking through the conditions and assumptions of one's own discipline; and, in particular, to the consequences for both leftists and liberals of not being able to give up their devotion to previously held notions of politics, progress, morality, sovereignty and so forth. Significantly, theory has been a regular target for moralists, Brown observes, frequently being chastised for its "failure" to tell the left what to struggle for and how to act (2001: 29). Indeed, Brown asserts that "moralism so loathes overt manifestations of power ... that the moralist inevitably feels antipathy toward politics as a domain of open contestation for power and hegemony"; and that "the identity of the moralist is," in fact, actually "staked against intellectual questioning that might dismantle the foundations of its own premises; its survival is imperiled by the very practice of open-ended intellectual inquiry."[50]

Bruining likewise draws on Brown's thinking on moralism (in his case, under the explicit influence of Joanna Zylinska's chapter on ethics in *New Cultural Studies*). He does so to show how, in the new materialist works he engages with, the emphasis on the concept of materiality, which in such discourses comes to represent that "universal and indisputable good that must be preserved," and criticism of poststructuralism and those modes of thought associated with it for not theorizing the material, is actually a form of reactionary "material foundationalism."[51] But just as interesting is the way such moralizing—also evident in the calls Braidotti associates with theory-fatigued neocommunist philosophers such as Badiou and Žižek to "return to concrete political action, even violent antagonism if necessary, rather than indulge in more theoretical speculations"—often takes the place of and in fact supplants genuine critical engagement.[52] In fact, Brown argues:

> Despite its righteous insistence on knowing what is True, Valuable, or Important, moralism as a hegemonic form of political expression, a dominant political sensibility, actually marks both analytic impotence and political aimlessness—a misrecognition of the political logics now organizing the world, a concomitant failure to discern any direction for action, and the loss of a clear object of political desire. In particular, the moralizing injunction to act, the contemporary academic formulation of political action as an imperative, might be read as a symptom of political paralysis in the face of radical political disorientation and as a kind of hysterical mask for the despair that attends such paralysis. ... Indeed, paralysis of this sort leads to far more than an experience of mere frustration: it paradoxically evinces precisely the nihilism, the antilife bearing, that it moralizes against in its nemesis—whether that nemesis is called conservatism, the forces of reaction, racism, postmodernism, or theory.[53]

Along with the emphasis on creative affirmation rather than negative critique, the anti-intellectualism of such moralism goes a long way toward explaining why many self-proclaimed new materialists so often indulge in the unthinking repetition of reductive clichés about theory in general and deconstruction in particular. What is more, it helps explain why they do so:

1. Without feeling the need to provide a careful, rigorous reading of specific thinkers and texts. (Unfortunately, Braidotti does not read Derrida's works in any detail in *The Posthuman*. Instead, the issue of what deconstruction is, is both decided in advance and excluded from her analysis.)
2. When an actual rigorous and responsible engagement with Derrida's philosophy would reveal that writing is nothing at all if it is not a material practice, even in the most obvious, received sense of the term. After all, a written mark, for it to be capable of being understood, must have a sense of permanence. This means it must be possible for it to be materially or empirically inscribed. In short, the condition of writing's very possibility is the material. This is why the transcendental is always impure, according to Derrida. Textuality and materiality, transcendence and immanence, even deconstruction and software code, as Federica Frabetti shows in *Software Theory*, cannot be set up in a dualistic relation in this respect as language and (theoretical) writing are already material.[54] We can thus see that deconstruction is much less a part of any supposed "linguistic turn" and is far more concerned with the material than it is portrayed as being in what, to borrow Braidotti's language, might be called zombie theories of materialism.

From Materialism to Materials

I mention all this not out of some stubborn refusal to move on and get with the new materialist program, or insistence on defending the legacy of Derrida and deconstruction— and even, if dare I say it, that this tradition needs to be reengaged "properly." As one of the cofounders of Open Humanities Press, I am partly responsible for the publication of at least two book series that can in their different ways be said to share Braidotti's impatience with deconstruction (Graham Harman and Bruno Latour's New Metaphysics and Tom Cohen and Claire Colebrook's Critical Climate Change).[55] Nor should any of the above be taken as implying we should ignore the question of what it means for our ways of being and acting as theorists if writing, print-on-paper, and the codex book are not regarded as the natural or normative media in which theoretical research and scholarship should be conducted (a question that can be said to represent part of posthumanism's own repressed, as we have seen); that we should instead continue to concentrate on writing linearly structured, original, fixed, and final print-on-paper texts, in uniform multiple-copy editions, on an all-rights-reserved basis, on the grounds that these too can be considered to be intricately bound up with the material. One of the reasons I am interested in Derrida—together with Bergson, Deleuze, Braidotti et al.—is that his philosophy has the potential to help those of us who are not resistant to thinking through the conditions and assumptions of our own disciplines (such as those that do indeed have to do with writing, print-on-paper, and the codex), and who are open to denaturalizing and destabilizing disciplinary formations (including formations associated with theory), to avoid slipping into such antipolitical moralism ourselves. Derrida's philosophy can do so by virtue of the way it teaches us to read and (re)write texts hospitably and responsibly. It also reminds us to be aware of how arguments such as those around materiality (and the proclaimed need to focus on software, hardware, life, biology, genetics, ecology, geology, the environment, electronic waste, and so on) are more often than not conducted in the very language and writing that such new materialism is ostensibly trying to move on from. A still further way Derrida's philosophy can assist us in avoiding the moralism that can be detected in many theories of materialism is through the emphasis it places on paying close attention not only to reading, writing, language, and the text—and with them, as we shall see, to the critical and creative rethinking of concepts such as precisely writing, language, text, materiality, matter—but to their material properties, practices, and processes of production. Hence the care Derrida devotes to considerations of the stylus or writing instrument;[56] the pen, the typewriter (first mechanical then electric), and word processor;[57] the signature;[58] paper;[59] the letter, postcard, "mystic wax writing pad,"[60] the book,[61] and of course its "inside matter";[62] as well as to the institutions of literature, philosophy, and the university.[63] (And that is without mentioning Derrida's many creative experiments with the material form, format, size, and shape of his texts: his use of multiple columns, extended footnotes, and so forth in works such as *Dissemination, The Post Card, Glas,* "Tympan," and "Circumfession.")[64]

It thus comes as no surprise to discover that not all materialists have succumbed to the tendency of zombie theory to position deconstruction in terms of the limitations of its concern with text, signification, and the linguistic. In *Quantum Anthropologies*, Vicki Kirby takes a highly sophisticated "materialist" approach to reading the legacy of Derrida in order to "recast the question of the anthropological—the human—in a more profound and destabilizing way than its disciplinary frame will allow."[65] Kirby proceeds to criticize as naive, complacent, and lacking in rigor John Protevi's claim in *Political Physics* that Derrida's "general text ... while inextricably binding force and signification in 'making sense', is not an engagement with matter itself." In fact, an engagement of this nature is not possible, if Protevi is to be believed, since matter, for deconstruction, "remains a concept, a philosopheme to be read in the text of metaphysics," or else "functions as a marker of a radical alterity outside the oppositions that make up the text of metaphysics."[66] Kirby condemns this account of deconstruction on Protevi's part on the grounds that the assumption that "Derrida's 'no outside metaphysics' *must* exclude matter ... entirely misses the extraordinary puzzle of how a system's apparent interiority can incorporate what appears to be separate and different." As far as she is concerned, while "'text' and 'metaphysics' are sites of excavation, discovery, and reinvention for Derrida, Protevi uncritically embraces their received meanings" (x).

It is also worth pointing out that not all Deleuzians have taken a moralistic approach to materialism. If "there are many Derridas," there are many Deleuzes too.[67] One of the most interesting articulations of a materialist ontology inspired by Bergson and Deleuze is that provided by another anthropologist, Tim Ingold. In an essay titled "Materials against Materiality," Ingold notes how despite the impression it gives, "the ever-growing literature ... that deals explicitly with the subjects of *materiality* and *material culture* seems to have hardly anything to say about *materials*": about the very matter of bodies, nonhuman objects, and the environment.[68] Instead, the concern of those emphasizing the importance of materiality is primarily with the language and writing of other theorists. The materials meanwhile have gone missing. As his title suggests, Ingold recommends a shift in attention from "the materiality of objects" precisely to "the properties of materials" (29). It is a move that, once made, is capable of exposing

> a tangled web of meandrine complexity, in which—among a myriad of other things—the secretions of gall wasps get caught up with old iron, acacia sap, goose feathers and calf-skins, and the residue from heated limestone mixes with emissions from pigs, cattle, hens and bees. For materials such as these do not present themselves as tokens of some common essence—materiality—that endows every worldly entity with its "objectness"; rather, they partake in the very processes of the world's ongoing generation and regeneration, of which things such as manuscripts or housefronts are impermanent by-products. (26)

All this goes at least some way toward explaining why I do not want to oppose the tradition of Spinoza, Bergson, and Deleuze and Guattari to that of Derrida, as from here it is not hard to see how even a rigorous reading of Deleuze's philosophy can be employed to show that deconstruction is actually far more concerned with matter, materials, and material factors than a lot of erstwhile (new) materialism. In fact, according to Ingold, instead of helping to theorize the material, the concept of materiality as it features in many studies of material culture serves to reinforce, rather than overcome, the classical (Latour would no doubt call them *modern*) dualities and dialectical oppositions between nature and culture, immaterial and material, language and reality, things and words, body and mind. This is because a part of the material world such as a rock or stone tends to be considered by discourses on materiality as "both a lump of matter that can be analysed for its physical properties and an object whose significance is drawn from its incorporation into the context of human affairs" (31). For Ingold, however,

> humans figure as much within the context for stones as do stones within the context for humans. And these contexts, far from lying on disparate levels of being, respectively social and natural, are established as overlapping regions of the *same* world. It is not as though this world were one of brute physicality, of mere matter, until people appeared on the scene to give it form and meaning. Stones, too, have histories, forged in ongoing relations with surroundings that may or may not include human beings and much else besides. (31)

He thus takes great care to distinguish between the "material world" of material culture theorists and a "world of materials" or, better, the *environment*. The material world exists in and for itself. The environment, by contrast,

> is a world that continually *unfolds* in relation to the beings that make a living there. ... And as the environment unfolds, so the materials of which it is comprised do not *exist*—like the objects of the material world—but *occur*. Thus the properties of materials, regarded as constituents of an environment, cannot be identified as fixed, essential attributes of things, but are rather processual and relational. They are neither objectively determined nor subjectively imagined but practically experienced. In that sense, every property is a condensed story. To describe the properties of materials is to tell the stories of what happens to them as they flow, mix and mutate. (30)

Performative Materiality and Media Archaeology

That said, if we are not to replicate the problems Ingold is identifying in the anthropological and archaeological literature on material culture—that it has surprisingly little to say about materials; that for all its rhetorical emphasis on the material, this literature is more involved with language, writing, theory, and philosophy; and that ultimately it reinforces, rather than challenges, the polarity between the material and immaterial, nature and culture—we need to exercise great care when it comes to bringing a processual and relational analysis of this kind to bear on approaches associated with the materialist turn in the humanities: neomaterialism, media archaeology, object-oriented philosophy, speculative realism, and so on.[69] Let me take, as a brief example with which to illustrate some of the difficulties inherent in doing so, Johanna Drucker's idea that the meaning and value of objects result from a performative act of interpretation motivated by their material properties and capacities—especially as this theory of performative materiality is articulated in relation to a reading of media archaeology that emerges from two of her recent texts: "Understanding Media," and "Performative Materiality and Theoretical Approaches to Interface." (Media archaeology is of particular interest in this context because, as Drucker observes, "materiality not only matters in media archaeology"; it is "the very subject of study.")[70]

To return to Ingold's analysis for a moment, in "Materials against Materiality," he emphasizes how it is

> significant that studies of so-called material culture have focused overwhelmingly on processes of consumption rather than production. For such studies take as their starting point a world of objects that has, as it were, *already crystallized out* from the fluxes of materials and their transformations. At this point materials appear to vanish, swallowed up by the very objects to which they have given birth. ... Thenceforth it is the objects themselves that capture our attention, no longer the materials of which they are made. It is as though our material involvement begins only when the stucco has already hardened on the house-front or the ink already dried on the page. We see the building and not the plaster of its walls; the words and not the ink with which they were written. (26)

And, to be sure, it is not difficult to see how a similar conclusion could be reached with regard to both Friedrich Kittler's concern with the way the physical materiality of "real" consumer media objects such as the gramophone, film, and typewriter "shaped the very conceptions of literary forms and formats," as Drucker puts it in "Understanding Media" (Kittler being one of media archaeology's key influences, even if he never embraced the term himself), and the importance that has subsequently been attached to how the "grooves of a wax recording or vinyl record are conceived and understood as writing, thus embodying an epistemological model" and the reading of that "model from the physical artifact, rather than reading the artifact for what it contains."[71] The latter is a characteristic media archaeological position, according to Drucker. It is one she associates with theorists such as Wolfgang Ernst, Jussi Parikka, and Lisa Gitelman, who are following in Kittler's tracks by studying the "particular material nature" of media.[72] On this basis, it could be shown that what we are confronted with here are media archaeological studies that, to repeat Ingold's words, "take as their starting point a world of objects that has ... *already crystallized out* from the fluxes of materials and their transformations" and from which the materials do indeed seem to disappear, subsumed as they are by the "objects to which they have given birth."[73]

I should stress at this point that it is not my intention to imply Drucker has been influenced by Ingold's account of the "dead hand of materiality" when writing either of these two essays.[74] (If she has, she certainly does not refer to either him or his work.) I am simply taking her reading of media archaeology here as an example, for it seems to me that it is not very far away at all from the kind of analysis that might easily be produced if we were to try to apply Ingold's critique of materiality to those discourses in the humanities that are associated with new materialism. This can be seen from the way Drucker, in her essay "Performative Materiality," makes a move similar to that of Ingold when he distinguishes between, on the one hand, the objects of the material world of material culture theorists and, on the other, the properties of materials that are processual and relational and are regarded as constituents of an environment. In Drucker's case, the distinction is between media archaeology and its emphasis on the material attributes of objects, artifacts, and entities and a perspective that presents the materiality of media more in terms of instability and flux:

> Some of the media archaeology approaches to studying in an *archaeographology*, to use Wolfgang Ernst's term, reinscribe digital media in an entity-driven approach that is both literal (code as inscription) and virtual (code as model) (Parikka, 2011).[75] These counteract the model of *immateriality*, though they do not replace it with a concept of digital flux, or of material as an illusion of stability constituted across instabilities.[76]

It is certainly tempting to view Drucker's own efforts to extend an ontological understanding of material things based on their properties and capacities with a performative dimension that "suggests that what something *is* has to be understood in terms of what it *does*, how it works within machinic, systemic, and cultural domains," as being capable of leading to a far more subtle and nuanced theory of materiality and materials than media archaeology's entity-driven approach—and all the more so given that Drucker makes the above point concerning media archaeology in the context of a larger argument for the digital humanities to reengage with the mainstream principles of critical theory. It is on the latter intellectual tradition that she bases her model of performative materiality, and, significantly, she includes in it not just structuralism and cultural studies, but poststructuralism and deconstruction as well. Yet things are not quite so simple. (The situation cannot be set up in terms of "new materialism of media archaeology bad," "Drucker's performative materiality and re-engagement with post-structuralist theory good.") It is by no means certain that Drucker's theory of performative media contests the dichotomy between the immaterial and material any more than does media archaeology on her account.[77] "Objects exist in the world," Drucker writes, "but their meaning and value are the result of a performative act of interpretation provoked by their specific qualities." Yet where do these performative acts and events originate? Are they ontologically distinct from material objects? Materiality clearly "provokes the performance." What is less clear on this basis is whether she considers the performance itself to be material or whether the performance transcends the material.

Is what we are presented with by Drucker in the guise of her theory of performative media merely another case of incorporeal, immaterial minds and their interpretative processes existing in a binary relation—albeit a dynamic one—with the material world, its objects, their qualities and properties?[78] It is a difficult question to answer, certainly on this evidence.[79] Still, I hope I have shown why a great deal of care does need to be taken when applying a processual and relational analysis of this nature to those approaches associated with the materialist turn in the humanities. As Drucker's account of media archaeology and theory of performative materiality illustrates, there is a significant risk in doing so of repeating the problems Ingold identifies in the anthropological and archaeological literature on material culture. We can thus see that determining the extent to which what Ingold reveals about the study of materiality is or is not the case with regard to media archaeology, neomaterialism, object-oriented philosophy, and speculative realism—or indeed performative materialism—is something that requires a careful, rigorous, singular, and even *performative* engagement with particular thinkers and texts.[80] It is something of this kind I am attempting to accomplish with regard to my reading of Braidotti's 2013 book published by Polity, *The Posthuman*.

"The Words and Not the Ink"

One thing it is perhaps fair to say at this point is that for all her emphasis on materiality, materialism, and grounding her rethinking of posthuman subjectivity in "real-life, world-historical conditions"—indeed, it could now be suggested precisely because of it—Braidotti says nothing about the ink with which *The Posthuman* is printed, to draw on an example of Ingold's (83). Nor, for that matter, does Braidotti concern herself with the ink with which she signed the contract with Polity that gives her the legal right, in accordance with the post-Enlightenment, postindustrial, advanced capitalist model of the individual human "subject as citizen, rights-holder, property-holder," to assert her identity as the "author of the work," and claim the original, fixed, and final version of *The Posthuman* as her isolable intellectual property.[81] In fact, if the examples I provided earlier in relation to Derrida's creative experiments with materials are anything to go by, Braidotti pays careful attention to the properties of very few of the actual materials that make up the real-life environment in which she is located and in which her knowledge and research come into being here and is performed, organized, categorized, published, exchanged, circulated, read, and engaged. I am thinking, in particular, of the paper and card on which her text is printed, its binding and glue (which can make books difficult to recycle), along with its design, format, size, layout, dimensions, font, typography, and so forth.

Granted, this could be a case of Braidotti being concerned with matter, materiality, and indeed "matter-reality" conceived more in terms of the world's continual unfolding and endlessly variable processes of change, flow, and mutation than as a constant, inherent property of objects or things that can be forensically and formally analyzed in this manner.[82] Following Ingold, we could think of this in terms of the "involvement" of *The Posthuman* in the "total surroundings" of its location, the "manifold ways in which it is engaged in the currents of the lifeworld."[83] Yet neither does Braidotti devote time to tracing any of the multiple currents of materials and their transformations that can be said to actively constitute the "world-in-formation" with which her practices and processes of production as a posthuman critical theorist are caught up, and of which a physical entity such as an ink-on-paper-and-card book bound together by glue is but an unstable and temporary "by-product," even if it does have the illusion of stability and permanence.[84]

It is not only when measured against criteria derived from Ingold's work that Braidotti does not appear to be especially materialist; this is also the case when it comes to her own criteria too. In her conclusion, Braidotti claims to have adopted "the speaking stance and the writing position of a tracker or cartographer" throughout *The Posthuman*, cartographic accuracy being one of her golden rules for posthuman theory, as we saw earlier (187). To this end Braidotti acknowledges that she is a "post-industrial subject of so-called advanced capitalism" occupying a "situated position as a female of the species." Yet she is also a subject who does not identify with the prevailing categories of subjectivity as a result of her "feminist consciousness of what it means to be embodied female" (80). And to be sure, the subject and self is extremely important for Braidotti. We need "at least *some* subject position" as the "site for political and ethical accountability, for collective imaginaries and shared aspirations," she insists (102). (This is why *The Posthuman* places so much emphasis on the issue of subjectivity, as in its own way does *Pirate Philosophy*.)

At the same time Braidotti emphasizes that her view of posthuman thought is "profoundly anti-individualistic" (101)—and all the more so given she considers being posthuman to imply

> a new way of combining ethical values with the well-being of an enlarged sense of community, which includes one's territorial or environmental inter-connections. This is an ethical bond of an altogether different sort from the self-interests of an individual subject, as defined along the canonical lines of classical humanism. ... Posthuman theory also bases the ethical relation on positive grounds of joint projects and activities, not on the negative or reactive grounds of shared vulnerability. (190)

In fact, her nomadic thought, which she insists is "rigorously materialist," goes to great lengths to champion a "postindividualistic" understanding of the subject (87). As far as she is concerned, a "sustainable ethics for non-unitary subjects" depends on an expanded and relational "sense of inter-connection between self-and others" (190), with subjectivity being very much an "assemblage that includes non-human agents" (82). Accordingly, the test this represents for critical theory is considerable, not to say huge: "We need to visualize the subject as a transversal entity encompassing the human, our genetic neighbors the animals and the earth as a whole" (82).

By taking Braidotti at her word when it comes to her argument about the importance of adopting an ethical, anti-individualistic approach, something else also becomes clear, however. The construct "Braidotti-the-internationally-respected-critical-theorist"—and with it Braidotti's identity as the sole, original, attributed author of *The Posthuman*—can be created and maintained only by not acknowledging, indeed only by marginalizing, repressing, ignoring, or excluding, those "ties that bind" Braidotti's own self to "multiple 'others' in a vital web of complex interactions," and which anchor her as a subject "in an ethical bond to alterity, to the multiple and external others that are constitutive of that entity which, out of laziness and habit, we call the 'self'" (100). (We might call this a Braidotti-network, following Latour. Or, better still, a Braidotti meshwork, to adopt Ingold's terminology once again).[85] Included among these multiple others are all those distributed, heterogeneous humans, nonhumans, objects, nonobjects, and nonanthropomorphic elements that collectively contribute to the emergence and history of an ink-on-paper-and-card book that is published under the name "Rosi Braidotti." Most obviously they take in her caregivers, teachers, colleagues, students, and peers, together with the publishers, editors, peer reviewers, designers, copyeditors, proofreaders, printers, publicists, marketers, distributors, retailers, purchasers, and readers of her book. But they also include all the other "multiple connections and lines of interaction that necessarily connect the text to its many 'outsides'":[86] not least those associated with the labor involved (including that of the agency workers, packers, and so-called ambassadors in Amazon's "fulfillment centers"),[87] the financial investments made, the energy and resources used, the plants, minerals, dyes, oils, petroleum distillates, salts, compounds and pigments, the transport, shipping and container costs, the environmental impact, and so forth.[88] Yet if this is the case, then surely when it comes to her subject formation as an author and authorial persona, one of the first obstacles of "self-centered individualism" (190) that needs to be removed, or at least challenged, is indeed the "possessive individualism" inherent in her claim to be identified as "author of this work" in accordance with UK copyright law, and use of an all-rights-reserved license (184). It is this possessive individualism, after all, that helps to shape and control the sites and situations through which her book can and cannot move, and which make it difficult both for *The Posthuman* to be seen as a collective project of any kind and to be jointly shared with, and reused by, others as common property on a nonprofit basis. Without such a challenge, are we not back with what Braidotti identifies as the "spirit of contemporary capitalism"?

> Under the cover of individualism, fuelled by a quantitative range of consumer choices, that system effectively promotes uniformity and conformism to the dominant ideology. The perversity of advanced capitalism, and its undeniable success, consists in reattaching the potential for experimentation with new subject formations back to an overinflated notion of possessive individualism, tied to the profit principle. This is precisely the opposite direction from the non-profit experimentations with intensity, which I defend in my theory of posthuman subjectivity. (61)

This is what I meant when I wondered previously whether repetitions of the negative and reductive new orthodoxies about theory, and the (non)decisions Braidotti makes about nonprofit, open source, open science, and collective ways of working, together with her not pushing further in the direction she herself sets us on toward the extreme of becoming posthumanities, are in some senses connected. Over and above the fact that it is impossible to escape the "constitutive blindness," the "contingency and selectivity" that "makes knowledge possible," by achieving perfect self-consciousness regarding our ways of acting and thinking as theorists, is it the attempt to move on from what is positioned as the now unfashionable emphasis on language, writing, and the text of the philosophical theories of poststructuralism and deconstruction that at least in part leads Braidotti to pay insufficiently rigorous attention to her own material reality as a theorist and philosopher here?[89] Does this explain why she appears to display a certain inertia with regard to her own "established mental habits" (58), practices, and processes, for all the emphasis that is placed in *The Posthuman* on producing a "radical estrangement" from the "normatively neutral relational structures of both subject formation and of possible ethical relations" (92). Is what is missing precisely the kind of qualitative shift Braidotti looks to when she states:

> A qualitative step forward is necessary if we want subjectivity to escape the regime of commodification that is the trait of our historical era, and experiment with virtual possibilities. We need to become the sorts of subjects who actively desire to reinvent subjectivity as a set of mutant values and to draw our pleasure from that, not from the perpetuation of familiar regimes. (93)

I hope all this explains still further why I do not want to oppose Deleuze to Derrida, and marginalize or forget about the likes of deconstruction and the attention it has paid to our ways of being as theorists and philosophers. If anything, discourses influenced by Derrida and his interest in (the material practices of) writing and language can help us think through this relation between mind and matter, often more productively than many of those associated with Deleuze who claim explicitly to be materialist.[90] At the very least, I want to hold in a complex tension—one that is perhaps better thought of in terms of contamination or diffraction, given the extent to which each is capable of transforming the identity of the other—the emphasis on a processual and relational political ontology, developed via that tradition of thought associated with Spinoza, Bergson, Deleuze, and Guattari, right through to contemporary theorists such as Jane Bennett, Steven Shaviro, and Braidotti herself, with the emphasis on the responsible decision that is more closely associated with the philosophical tradition running from at least Emmanuel Lévinas and Derrida, to take in contemporary philosophers including Ernesto Laclau and Chantal Mouffe—for whom politics is a decision taken in an undecidable terrain, very much as it is for

Derrida—and, building on the work of Karen Barad and her notion of the "agential cut," Sarah Kember, and Joanna Zylinska.[91]

From this perspective, what is helpful about the latter tradition's account of the undecidable nature of the decision is that it enables us to understand and rigorously think through how any such decision necessarily involves a moment of madness. This is important, because once we appreciate that the decision is the invention of the other, including the other in us (i.e., the environment, animals, insects, plants, dust, technology, the planet, the cosmos), and thus is not of complete human control, we can endeavor to take on—or, perhaps better, endure "in a *passion*"—rather than simply act out, take for granted, forget, repress, ignore, or otherwise marginalize the implications of this realization for the way we live, work, act, and think as theorists.[92] This includes what actions to take (or not to take in the case of strategic withdrawal into inactivity, silence, and passive sabotage—see chapter 5), what readings or writings to produce (concerning materialism and so forth), even what affirmative and creative flows of desire to follow. We can do so in an effort to make the impossible decisions that confront us (e.g., those concerning theory; nonprofit; collectivity; open source; open science; open access; the unified, sovereign, proprietorial subject; the individualized author; copyright; and so on) as responsibly, and with as much care and thought, as possible.

Part 4: Posthumanities

That being the case, there are a number of perfectly understandable reasons why Braidotti, Wolfe, and others associated with the posthuman—or the "posthumanist" to be strictly correct in Wolfe's case[93]—might have nonetheless taken a decision not to create, publish, and circulate their research in a collective, free, and open manner; why, for the most part, they have instead adhered more closely to the established forms and methods of disciplinary knowledge of the humanities than the kind of rules, guidelines, and criteria Braidotti lays out for posthuman critical theory. What is more, these reasons may also help to explain why they have done so for all their critique of essentialist, humanist notions of identity and subjectivity, and insistence that the "point of critical theory is to upset common opinion (*doxa*), not to confirm it."[94] For as Braidotti observes in *The Posthuman* with regard to another of her methodological guidelines for posthuman theory, disidentification (e.g., from the dominant institutions and their normative representations of class, gender, race, sexuality, age, authorship, originality, of how particular persons and positions of power and influence emerge, and therefore of what it is to be an internationally-respected critical theorist), at the same time as it can pave the way for "creative alternatives" (89), can also be highly destabilizing, as it "involves the loss of cherished habits of thought and representation, a move which can also produce fear, sense of insecurity and nostalgia" (168). When it comes to methodology, "it requires dis-identification from century-old habits of anthropocentric thought and humanist arrogance"—including, I would argue, those that are associated with monographs, journals, journal articles, conferences, research centers, protocols of advancement and recognition, etc.—"which is likely to test the ability and willingness of the Humanities" (168).[95]

To stay with the example of subjectivity, individualism, and copyright, Braidotti tells us that the alternative processes of becoming that conceptual personae such as the feminist, the queer, and the cyborg dramatize "defy the established modes of theoretical representation, because they are zigzagging, not linear and process-oriented, not concept-driven" (164). Yet it can be argued that if the posthuman subject is conceived in terms of a relational and multifaceted process of becoming—understood, following Deleuze and Guattari along with Braidotti's favorite philosopher, Spinoza, as being part of a monistic ontology—then the work and thought of any correspondingly zigzagging, rhizomatic, nonlinear, process-oriented methodology or way of becoming adopted by posthuman critical theorists would similarly defy the established modes of applying intellectual property laws and asserting copyright, certainly on an all-rights-reserved basis. This is due to a distinction that is made in our political economy—not every political economy, but our particular political economy—between the "process of making" and the "finished object that is made." As the anthropologist James Leach emphasizes in "The Politics of (Making) Knowledge Objects," it is not possible to "own a distinctive form of creative practice, only the expressions of that practice." This difference is vital to intellectual property law. It "amounts to the distinction between idea and expression, with the expression as that which can be protected. Under this logic, such protection is appropriate because it is the expression, not the idea or the process of making, which has the value (value creation in transaction determined by consumer market)."[96] As a consequence, policy and precedent focus too frequently "on an object and its value to the detriment of the processes whereby wider social value" and benefit are created (80). What is more, this is also the case with regard to knowledge created in universities. Here, too, a regularly repeated theme takes shape whereby the "emphasis for claims, for calculating recompense ... locates value ... not in the processes of production" (80). Value is located rather in the production of things—of discrete finished objects that are attached to, but externalized from, individuals and have qualities analogous to transactable commodities in that they can be abstracted from the social context and relations of production and their value relocated elsewhere. Books, for example, for which the academic "author of the work," as established in accordance with copyright law, can claim recompense and accrue value, recognition, advancement, credibility, prestige. These are still regarded as the key achievements.

Ideas such as the unified, sovereign, proprietorial subject, the individualized author, the signature, fixity, copyright, and so on may be a means of credentialing authority and creating trust between the author and reader, then.[97] But as Janneke Adema reveals in her research into the history of the scholarly book, building on that of the historian Adrian Johns, it was precisely the conventions related to propriety and trust, developed by publishers in the eighteenth century as protection against piracy and impropriety, that were also used to turn the book into a scholarly object and commercial commodity.[98] So to adopt a methodology that does indeed focus more on the multifaceted and relational processes of making and becoming would involve posthuman critical theory in throwing open a number of radical questions that would be profoundly destabilizing for the capitalist economy's commodification of knowledge and the way value here "lies in objects, sites, or codifiable (that is static) practices"—hence the emphasis in universities today on knowledge that can be objectified and thus measured, audited, used, transferred (e.g., to businesses in the form of knowledge transfer partnerships).[99] To be sure, these questions would be destabilizing for conventional ideas of open data and open access, since such research in posthuman critical theory would be understood not so much as an object to which free, gratis access can be given (and which in the case of the majority of even open access books, libre access apparently cannot), but as a process. Moreover, it would be a process that universities teach and to which, as we have seen with the development of open education and open science, free access can be given.[100] But such radical questions would also be profoundly destabilizing for many of the ideas on which the humanities are based and which shape how we live, work, and act as theorists, including how our subjectivities as (in some cases, internationally respected) philosophers are created and maintained.

None of this is to ignore or deny that research in critical theory cannot only be processual and has to be an academic or commodifiable object, site, or practice to some degree, not least if we are to make it available on an open access, open knowledge, or open education basis, or continue to operate effectively within the institution of the university. (More or less violent decisions, cuts, or interruptions do have to be made.) One can imagine part of Braidotti's aim in writing this somewhat more accessible book is to open up her theories on the posthuman, on advanced postindustrial capitalism, on feminism, and on nomadic ethics, to a wider audience both within the institution and without—in which case, to have actually assumed or endured in a passion her ideas about nonprofitability, the collective, open source, and so on might well have put this strategy at risk. It would have required her to act, think, and write somewhat differently as a theorist from how the construct known as "Braidotti-the-internationally-respected-critical-theorist" is created and maintained. Publishing *The Posthuman* on such a basis might not have had the same status as publishing "all rights reserved" with the for-profit print publisher Polity. It would thus not have fit quite so well with the protocols of recognition and credibility in the humanities and with how one achieves power and influence, and thus often has an effect and impact.[101] Indeed, we have to work strategically in particular contingent contexts and make the best—or least worst—decisions and cuts possible, which is why I would not want what I have argued here to be taken as a moralistic critique of Braidotti—or Wolfe—on my part. (After all, it is from the conventional, "Copyright ... all rights reserved," print books of Braidotti, Wolfe, and others like them that I have learned much of this.)[102]

Yet the above does highlight some of the (non)decisions that have been made (not deciding to do anything about publishing on a nonprofit, collective basis also being a decision). This is another reason I am focusing on Braidotti's book here: because it makes these (non)decisions relatively clear. A still further reason I am concentrating on *The Posthuman* is because of the way it draws attention to some of the other, perhaps more responsible and caring decisions, cuts, actions, and interruptions that are possible. What is interesting about them is they are interventions that may enable us as theorists to both critique certain forms of humanism and the humanities and embark, creatively and constructively, on some alternative ways of being. They may also help generate relationships between persons, nonhumans, and objects we might begin to think of in terms of the possibility of becoming posthumanities, without being pulled back to humanism and the humanities.

In this respect, any such posthumanities would not represent an attempt to radically "overcome Humanism as an intellectual tradition, a normative frame and an institutionalized practice" (30). The posthumanities would rather be more of a mutation or intensification of elements, dynamics, tendencies, and potentials already present and preinscribed in the humanities, and which challenge their underpinning humanism, including those forms of humanistic knowledge and subjectivity associated with books, monographs, journals, "the authority of a proper noun, a signature, a canon, a tradition, or the prestige of an academic discipline" (165). (This is why there is *not* simply a humanism for posthumanism to come after and why there *can* be such a thing as "posthumanist Shakespeare," for example, as posthumanism on this account is not confined to the modern era, with its twentieth- and twenty-first-century robots, reproductive technologies, neuromedicine, and genetically modified food. Humanism is in this sense already posthumanist.) Far from a dialectical attempt to move on from the human and the humanities by announcing their end, the posthumanities would call for even greater care and attention to be paid to the former. (To do otherwise, as we have seen, risks the unwitting re-creation of the same old narratives, patterns, tropes, and structures of analysis, albeit perhaps in a different guise.) I hope that this conception of the posthumanities as a mutation or intensification also provides a justification for my arguing for all this by means of a careful critical engagement with the humanism and (posthuman) humanities of Braidotti's for-profit, all-rights-reserved monograph, and a somewhat irreverent teasing and testing of its limits.

Postmonograph

So it is not an either-or choice of humanities or posthumanities. As Open Humanities Press and a number of other open access publishers demonstrate, one can publish a quite traditional, linearly structured book-length study and still make it freely available on a nonprofit basis using open source software and Creative Commons licenses. Some are even finding ways of doing so with a book published by a brand-name, for-profit, all-rights-reserving print-on-paper-only press, such as by circulating a "pirate" copy on a free text-sharing network like Aaaaarg, Monoskop, or the Library Genesis Project's LibGen.[103] This is why, to Braidotti's alternative figurations of the feminist, the queer, the cyborg, the nomadic subject, and so on, I want to add a conceptual persona akin to that of the pirate—although (as I make clear in chapter 1) when I use the word *pirate*, I have in mind chiefly the etymological origins of the modern term with the ancient Greek noun *peira* and verb *peiraō*. The pirate here is someone who makes an attempt, tries, teases, troubles, gets experience of, endeavors, attacks.[104] As examples such as Aaaaarg (which has been experimenting with forms of annotated biography not so very far away from the Living Books About Life series) and The Piracy Project show,[105] art offers one space where assumptions about property and propriety can be contended with and tested, by means of an emphasis on nonprofit, collectivity, and process especially.[106] Thanks in part to developments in electronic publishing, open education, and the digital humanities, however, even the more conventional disciplinary formations of critical theory and philosophy, heavily intertwined with the commodified print-on-paper codex text though they may be, have the potential to offer such a space too.[107] (And I maintain this is the case in spite of the way many philosophers today appear intent on returning us to a period of "phallogocentric emphasis," as Hélène Cixous describes it, "heavily masculine and devoid of imagination.")[108]

Now that open access has apparently reached a tipping point, with approximately half of the scientific papers published in 2011 available for free, could we be on the verge of a postmonograph era?[109] It is certainly a direction the artist and poet Kenneth Goldsmith is pointing us in. Goldsmith's argument is that with the development of the Web, writing has encountered its photography. What he means by this is that it has come up against a "situation similar to what happened to painting with the invention of photography, a technology so much better at replicating reality that, in order to survive, painting had to alter its course radically." Writing's most likely response, according to Goldsmith, is to be

> mimetic and replicative, primarily involving methods of distribution, while proposing new platforms of receivership and readership. Words very well might not only be written to be read but rather to be shared, moved, and manipulated, sometimes by humans, more often by machines, providing us with an extraordinary opportunity to reconsider what writing is and to define new roles for the writer. While traditional notions of writing are primarily focused on "originality" and "creativity," the digital environment fosters new skill sets that include "manipulation" and "management" of the heaps of already existent and ever-increasing language.[110]

Of course Goldsmith is ultimately looking for new ways to arrive at what are quite conventional goals—or, as he puts it, "unexpected ways to create works that are as expressive and meaningful as works constructed in more traditional ways."[111] What, though, if we wish to place in question the humanist values associated with being creative, expressive, meaningful, and indeed with writing? Just as Goldsmith has been criticized for being too dialectical because he fails to question creativity and simply opposes it, could we not say something similar about his continued emphasis on writing?[112] At this moment in time, it is difficult to think of an important continental philosopher who does not focus in the main on writing papercentric books, book chapters, and journal articles. No doubt this is because the university has historically been an institutional expression of the book.[113]

What those of us associated with the Living Books About Life project are interested in, however, is what it will mean for our ways of working and thinking as theorists if writing, print-on-paper, and the codex book are no longer necessarily held as the natural or normative media in which research and scholarship are to be conducted and distributed in the humanities, but are rather absorbed into a variety of other, often multimodal and hybrid forms of communication, including those made possible by networked computers, cameras, smart phones, tablets, databases, archives, wikis, and file-sharing networks?[114] I mean by this research and scholarship that is not just communicated, disseminated, or promoted using this new media—or (re)configured into a digital form at some point in the composition, editing, and printing process (e.g., into a software package such as Microsoft Word or Adobe InDesign)—but is rather "born" with it. (I use the term *born* in the sense that the norms and practices of creation, presentation, and critical attention of this research and scholarship are associated with these new media rather than simply being those that have been inherited from the era of writing, the book, and the industrialization of printing.) This is not to suggest that writing and language can or should be abandoned. In fact, as Goldsmith points out, "even as the digital revolution grows more imagistic and motion-based (propelled by language), there's been a huge increase in text-based forms, from typing e-mails to writing blog posts, text messaging, social networking status updates, and Twitter blasts: we're deeper in words than we've ever been."[115] Again, rather than attempting to dialectically come after writing and the book by announcing their end, we see this posthumanities interest in other, often hybrid forms of communication, as enabling us to pay even greater attention to them.

The argument presented thus far constitutes a critically and creatively remixed, zigzagging, provocative way of explaining why those of us associated with Open Humanities Press and the Living Books About Life project are prepared to put many of the ideas on which the humanities are based at risk by making the books in the series available on an open read-write-rewrite basis. In keeping with this pirate philosophy and under the guise of what can be thought of as digital posthumanities, we are trying to test, trial, tease, and trouble some of these ideas rather than being continually dragged back toward the humanities, digital humanities, or even a posthuman humanities studies. As well as repackaging the available open access science material on life into a series of books, we thus see the Living Books about Life series as engaging in rethinking the book itself as a living, collaborative, nonprofit, processual endeavor. After all, as Braidotti writes: "Only a serious mutation can ... help the Humanities to grow out of some of their entrenched bad habits" (153).

I hope this chapter has explained why, when it comes to Open Humanities Press, we are experimenting with:

- Working on a nonprofit basis. All OHP books and journals are available open access on a free *gratis* basis and, as we have seen, many of them *libre* too.
- Operating as a radically heterogeneous collective of theorists, philosophers, scholars, librarians, publishers, technologists, journal editors, and so forth. OHP acts as a networked, cooperative, multiuser collective, where editors both internal and external to OHP "support one another and share knowledge and skills, very much like an open source software community."[116]
- Using open source software. Approximately half of OHP's journals use the Public Knowledge Project's Open Journal Systems software, one-quarter WordPress, with the remainder publishing on systems they have created themselves.[117]
- Gifting our labor rather than always insisting on being paid for it. We see this approach as a means of helping to decenter waged work from its privileged place in late capitalist neoliberal society and placing more emphasis on unwaged activities, including different kinds of carework.[118] But as we know from Roberto Esposito, the gift and giving also provide a means of developing notions of the Common and of community that break with the conditions supporting the unified, sovereign, proprietorial subject.[119]
- Working in a nonrivalrous, noncompetitive fashion to explore new models of ownership. OHP shares its knowledge, expertise, and even its publications freely with other open access presses, including Open Book Publishers at Cambridge, Open Edition in France, and both meson press and the Hybrid Publishing Lab at Leuphana University in Germany.[120]

That said, OHP continues to operate within certain limits—even if, with projects such as the Living Books About Life series, it is perhaps questioning and stretching those limits more than most, helping to give open access different inflections in the process. This aspect of OHP's working within the frame of the conventional and the given is especially apparent with regard to ideas of authorship, originality, and copyright. Both the press and the Living Books series continue to make use of Creative Commons licenses, for example, with all their attendant problems. (I mention this to emphasize that nothing I have written here is intended to convey the impression that the ways of being, thinking, and doing of Braidotti, Wolfe, et al. are the problem to which those of myself and my colleagues at Open Humanities Press are somehow the solution. As chapter 1 shows, Creative Commons is itself liberal and individualistic, offering authors a range of licenses from which they can personally choose rather than advancing a collective philosophy. The concern of Creative Commons is also with preserving the rights of copyright owners rather than with granting them to users, presuming everything that is created by an author or artist is his or her property.)

What is needed if this particular aspect of our ways of being theorists is to be challenged and changed are economic, legal, and political alternatives to publishing either on an all-rights-reserved copyright or open access and Creative Commons basis that are professionally recognized. We need new languages for doing so, new ways of living, working, acting, and thinking as critical theorists and radical philosophers. The chapter that follows explores some of the possibilities for developing such new languages and new ways of being. In particular, chapter 5 attempts to test, trial, and trouble such economic, legal, and political frames and limits by asking, What would it be for us, as theorists and philosophers, to adopt something approaching the figure or persona of the pirate—someone who traditionally has often operated in a manner that is neither simply legal nor illegal?

5 COPYRIGHT AND PIRACY
Pirate Radical Philosophy

The defining characteristics [of the posthuman] involve the construction of subjectivity, not the presence of non-biological components.[1]—N. Katherine Hayles

The "crisis of capitalism" and the associated series of events known for short as the Arab Spring, Occupy, and #GlobalRevolution have been addressed at length by many critical theorists and radical philosophers. The same, however, cannot quite be said about what followed: the fact that a neoliberal government returned to rule in a Spain so shaken by the 15M movement; the conservative Enrique Peña Nieto became president of Mexico regardless of the Yo Soy 132 protests; Recep Tayyip Erdoğan continued to lead a post-#DirenGezi Turkey "with an iron capitalist fist" even when his AKP party no longer enjoyed a parliamentary majority;[2] and the military were able to regain control of the Egypt that had overthrown Hosni Mubarak and thus derail any revolutionary dynamic, with a former general, Abdel Fattah el-Sisi, eventually becoming president.

But let us be generous and attribute this discursive shortfall largely to reasons of timing. After all, some of these developments are still relatively new (certainly in terms of the speed at which scholarly publishing operates). One question raised by the postcrash political protests remains nonetheless. This concerns the degree to which the contemporary sociopolitical situation, in its various iterations, also poses a challenge to those of us who work and study in the university—a challenge that encourages us to go further than endeavoring to "just say no" to the idea of universities operating as for-profit businesses in order to serve the market economy, and demanding a return to the kind of publicly financed mass education policy that prevailed in the Keynesian era. (Rather than bringing radical change to the institution, too often the latter demand appears designed merely to prevent a group of faculty and students from being denied future careers in the university as it currently stands.) Of course, we must avoid making the mistake of seeing in each event, or even series of events, a tipping point that is at last turning the situation to our advantage; or, equally, of pronouncing that the global revolution is over and the counterrevolution has won. Still, keeping these qualifications in mind—something it is even more important to do in the light of the experience of Syriza in Greece and growing popularity of Podemos and the wider *indignados* movement in Spain —what if we too, in our capacity as academics, authors, thinkers, and scholars, wish to resist the continued imposition of a neoliberal political rationality that may at times appear dead on its feet but is still managing to blunder on? Suppose we desire a very different university from the one we have, but have no wish to retain or restore the paternalistic, class-bound model associated with the writings of Matthew Arnold, F. R. Leavis, and Cardinal Newman? While appreciating the idea that there is an outside to the university is itself a university idea, and that attempts to move beyond the institution too often leave it in place and uncontested, is it possible nevertheless to derive impetus from the emergence of autonomous, self-organized learning communities such as The Public School and free text-sharing networks such as Aaaaarg?[3] Does the struggle against the *businessification* of the university not call on us too to have the courage to attempt new economic, legal, and political systems and models for the production, publication, sharing, and discussion of knowledge and ideas?

To date, such questions have proven surprisingly difficult to bring into focus, no doubt in part because of the potential they contain to change and renew, radically, our professional practices and identities. In the March immediately following the student protests of November 2010, the Institute of Contemporary Arts in London hosted an afternoon of talks under the title "Radical Publishing: What Are We Struggling For?"[4] At first sight, this event looked as if it was going to explore some of these issues. As it turned out, the afternoon featured extensive discussion from a variety of speakers including Franco "Bifo" Berardi, David Graeber, Peter Hallward, and Mark Fisher about politics understood according to the most easy-to-identify signs and labels, the majority of which concerned political transformation elsewhere: in the past, the future, Egypt. Somewhat surprisingly, given its title, there was very little discussion of anything that would actually affect the work, business, role, and practices of the speakers themselves: radical ideas of publishing with transformed modes of production, say. As a result, the event in the end risked appearing mainly to be about a few publishers, including Verso, Pluto, and Zero Books, that may indeed publish radical political content but in fact operate according to quite traditional business models (certainly when compared to some of those that are considered in this chapter and the one that follows), promoting their authors and products and providing more goods for the ticket-paying audience to buy. If the content of their publications is politically transformative, their publishing models certainly are not, with phenomena such as the student protests and ideas of communism all being turned into commodities to be marketed and sold.

The Human

Blind spots of this nature are widespread throughout the humanities (as we have seen in the preceding chapters of this book). Consider the very idea on which the humanities are based: that of the human itself. The humanities have interrogated the concept of the human for the last hundred years and more, not least in the guise of critical theory and continental philosophy. Yet (and as chapters 3 and 4 in particular both show) the dominant mode of production of knowledge and research in the humanities continues to be tied to the idea of the indivisible and individualized, human(ist) author. It is a description of how ideas and concepts are created, developed, and circulated that is as applicable to the latest generation of theorists to emerge as it is to the "golden generation" of Barthes, Foucault, Lyotard, and Lacan—not just radical philosophers such as Agamben, Badiou, Latour, or Stiegler, but many of the so-called children of the 68ers like Meillassoux too. For all that theorists today may be more inclined to write using a computer keyboard and screen than a fountain pen or typewriter, their way of creating, publishing, and disseminating theory and theoretical concepts remains much the same. This is the case with respect to the initial production of their texts and their materiality. Even if they *are* configured into a digital form at some point in the composition, editing, and printing process, the orientation is very much toward the generation of print-on-paper codex books and articles (or at the very least paper-centric texts), written by lone scholars usually in a study or office and designed to make a forceful, authoritative, masterly contribution to knowledge. But it is also the case with regard to the attribution of their texts to individualized human beings whose identities—regardless of any associations they may have with antihumanist, posthumanist, or postanthropocentric philosophy—are unified and self-present enough for them to be able to claim them as their original work or property. Thus, while they adhere to one part of Walter Benjamin's suggestion in "The Author as Producer" that authors should feed the production apparatus with revolutionary content, they continue to ignore the other part: that authors should also change the production apparatus itself.[5]

Admittedly, these traditional methods for the creation, composition, publication, and circulation of knowledge and research in the humanities are being brought into question by that emergent body of work known as the digital humanities. Thus we have literary theorist Stanley Fish's characterization of those forms of communication associated with the digital humanities, blogs especially, as "provisional, ephemeral, interactive, communal, available to challenge, interruption and interpolation." Fish positions such uses of networked digital media technologies as standing directly against the traditional ambition of the scholarly critic, an ambition he admits to sharing. This entails being able "to write about a topic with such force and completeness that no other critic will be able to say a word about it." It is an aim he ascribes to a "desire for pre-eminence, authority and disciplinary power." Accordingly, Fish contrasts both blogs and the digital humanities to the kind of "long-form scholarship—books and articles submitted to learned journals and university presses"—he has devoted his professional life to and which he describes in terms of the building of "arguments that are intended to be decisive, comprehensive, monumental, definitive and, most important, *all mine*."[6]

The first thing to point out here is that as a diverse constellation of fields, the digital humanities is neither unified nor self-identical. It comprises a wide range of often conflicting attitudes, approaches, and practices that are being negotiated and employed in a variety of different contexts. That said, one broad definition of the digital humanities has it embracing all those scholarly activities in the humanities that involve writing about digital media and technology and being engaged in processes of digital media production, practice, and analysis.[7] Examples are developing new media theory; creating interactive electronic literature and archives; building online journals, libraries, databases, and wikis; producing multimedia museums and art galleries; and exploring how various technologies and methods drawn from computer science and related fields, such as the mining, aggregation, management, and manipulation of data, are reshaping teaching, learning, research, and publication, including peer review. Within this, the digital humanities undoubtedly provide us with a chance (if we can only take it) to experiment with some of the new tools, techniques, methods, and materials that digital media technologies create and make possible in order to bring new forms of Foucauldian *dispositifs*—what Stiegler calls *hypomnémata* and Plato describes as *pharmaka*, neither simply poisons nor cures—into play. And this includes *dispositifs* and *pharmaka* associated with our systems of higher education. Nevertheless, the extent to which specific performances of the digital humanities actually succeed in contesting any of "the interdependent notions of author, text and originality" that Fish positions as being required by the traditional model of long-form scholarship requires careful scrutiny.[8]

Certainly the digital humanist whom Fish concentrates on in most detail, Kathleen Fitzpatrick, does not offer a profound challenge to ideas of the human, subjectivity, or the associated concept of the author at all. Nor, to be fair, is she particularly interested in doing so. In fact, far from questioning radically the notion of the human that underpins "the 'myth' of the stand-alone, masterful author" (and, indeed, the university and, within it, the humanities), Fitzpatrick's view of the digital humanities sees it as being more concerned with bringing the humanities as they are traditionally known and understood to bear on computing technologies.[9] Take her book *Planned Obsolescence*, which, as an experiment with open peer review, was itself first published on a blog others could contribute to. Fish portrays Fitzpatrick as contending in this volume,

> first, that authorship has never been thus isolated—one always writes against the background of, and in conversation with, innumerable predecessors and contemporaries who are in effect one's collaborators—and, second, that the "myth" of the stand-alone, masterful author is exposed for the fiction it is by the new forms of communication—blogs, links, hypertext, re-mixes, mash-ups, multi-modalities and much more—that have emerged with the development of digital technology.[10]

Yet as Fitzpatrick makes clear in a section expressly concerned with the change in authorship, "From Individual to Collaborative,"

> the kinds of collaboration I'm interested in need not necessarily result in literal co-authorship. ... The shift that I'm calling for may therefore be ... less a call necessarily for writing in groups than for a shift in our focus from the individualistic parts of our work to those that are more collective, more socially situated ... focusing on this social mode of conversation, rather than becoming obsessed with what we, unique individuals that we are, have to say, may produce better exchanges. One need not literally share authorship of one's texts in order to share the process of writing those texts themselves; the collaboration that digital publishing networks may inspire might parallel, for instance, the writing groups in which many scholars already share their work, seeking feedback while the work is in process.[11]

Fish reads this as suggesting that "if the individual is defined and constituted by relationships, the individual is not really an entity that can be said to have ownership of either its intentions or their effects; the individual is (as poststructuralist theory used to tell us) just a relay through which messages circulating in the network pass and are sent along."[12] As Fitzpatrick emphasizes, however, the shift she is calling for is "less radical than it initially sounds." Far from being based on a rigorous decentering of the subject, her approach often seems closer to the liberal-democratic humanist stance she is endeavoring to question. Albeit it is one in which "unique," stable, centered authors are now involved in a "social" conversation "composed of individuals" that is somewhat akin to Habermas's ideal speech situation—at least to the extent this "conversation" appears to contain relatively little conflict, antagonism, or incommensurability between the participants.[13] There is no *differend,* as Lyotard puts it. Responding to Fish on his own blog, Fitzpatrick is at pains to point out she is not maintaining that notions of the author, text, and originality "are going away in the digital age, only that they are changing, as the interpretive community of scholars changes."[14] And, to be sure, it remains the case that for all her emphasis on "texts published in a network environment" becoming "multi-author by virtue of their interpenetration with the writings of others," she herself very much retains authorial control of *Planned Obsolescence.* Fitzpatrick continues to be the clearly identifiable "original" author of this clearly recognizable text, and it is to her personally that this text is clearly to be attributed.[15]

In this respect, it is significant that Fitzpatrick chose to employ a blogging tool for her experiment with open peer review, namely, WordPress, albeit with the CommentPress plug-in developed by the Institute for the Future of the Book that enables comments to appear alongside the main body of the text on a paragraph-by-paragraph, whole-page, or entire-document basis. For of course most blogs (in contrast to wikis, say; see chapter 4) do not actually allow collaborative writing, let alone the "elimination of the individual." The work of a blog's author tends to be kept quite separate from that of others who use the same blog to review or respond to that work or enter into conversation and dialogue with it. Although "responses to the text" may indeed "appear in the same form, and the same frame, as the text itself" (albeit usually vertically below the line rather than horizontally alongside, as is the case with CommentPress and its attempt to subvert this spatial and conceptual hierarchy), these two distinct identities and roles—of original author and secondary reviewer, respondent, or commentator—are maintained and reinforced by the blogging medium.[16] So not only does

Fitzpatrick not actually put ideas of the human, subjectivity, or the associated concept of the author to the test, neither do blogs, for all Fish endeavors to portray both otherwise. Instead, the maintenance of authorship and originality on Fitzpatrick's part is achieved with the assistance of the very medium (blogging) Fish positions as creating problems for it.

Ironically, then, it turns out that Fish has a far more radical vision of the digital humanities than Fitzpatrick. In fact, contrary to Fish's presentation of them, the way the majority of academics interact with blogs and social networks such as Facebook and Google+ actually functions to promote and sustain notions of the author and originality more than they undermine them. This is in no small part due to the fact that, as Felix Stalder points out, you "have to present yourself in public as an individual in order to be able to join digital social networks, which, increasingly, becomes a precondition [to] join other forms of social networking."[17] Such personal social networks may thus be seen to offer a variation on the theme of what Beverley Skeggs terms "compulsory individuality"—with a lot of academics using them as a means of promoting and marketing themselves, their work and ideas, not least by gathering "friends" and "circles" to network with and presenting themselves as accessible, engaged, charismatic personalities who are *always on*.[18]

So where does this leave us, if even the digital humanities (or at least Fish and Fitzpatrick's versions of them) do not represent much of a test of the orthodox modes of creation, composition, legitimization, accreditation, publication, and dissemination in the humanities? In a book from 2009, one of the participants in the Institute of Contemporary Arts' Radical Publishing event, Franco "Bifo" Berardi, raises the question as to whether we should not "free ourselves from the thirst" for the kind of activism he sees as having become influential as a result of the anticapitalist globalization movement: "Isn't the path towards the autonomy of the social from economic and military mobilization only possible through a withdrawal into inactivity, silence, and passive sabotage?" he asks. In light of this question, should we consider embracing our own variation on the theme of refusal that has been so important to autonomous politics in Italy: a strategic withdrawal of our academic labor—and not just from blogs and corporate social networks such as Facebook and Google+?[19]

Open Access

In January 2012, Peter Suber, a leading voice in the open access movement, provided an instance of just such a withdrawal. He announced on Google+

that he would "not referee for a publisher belonging to the Association of American Publishers unless it has publicly disavowed the AAP's position on the Research Works Act." The latter legislation, introduced in the US Congress on December 16, 2011, was designed to prohibit open access mandates for federally funded research in the United States. It would thus in effect have countermanded the National Institutes of Health's public access policy along with other similar open access policies in the United States. To show my support for open access and Suber's initiative, I publicly stated in January 2012 that I would act similarly.[20] At the time, this meant not writing, publishing, editing, or peer reviewing for, among others, Sage (which publishes numerous journals in the critical theory area, including *Theory, Culture and Society* and *New Media and Society*), and Palgrave Macmillan (publisher of *Feminist Review*), as well as Stanford University Press, Fordham University Press, Harvard University Press, and New York University Press. Having met with staunch opposition from within both the academic and publishing communities in what some dubbed the Academic Spring, all public backing of the Research Works Act was dropped as of February 27, 2012. But rather than taking this as a cue to abandon the strategy of refusal, I cannot help wondering if we should not regard this small victory as encouragement and adopt it all the more. For example, should we not withdraw our labor from all those presses and journals that do not allow authors, as a bare minimum, to self-archive the refereed and accepted final drafts of their publications in institutional open access repositories and use this time to become actively involved in the process of publishing open access on a nonprofit, free to both read *and* publish, basis instead?[21]

As a longstanding supporter, I believe it is important to acknowledge that the open access movement, which is concerned with making peer-reviewed research literature freely available online to all those able to access the Internet, is hardly any more unified or self-identical than the digital humanities. Some regard it as a movement, while for others, it represents merely a variety of economic models or even just another means of distribution, marketing, and promotion.[22] It should also be borne in mind that there is nothing inherently radical, emancipatory, oppositional, or even politically or culturally progressive about open access. The politics of open access depends at the very least on the approach adopted (e.g., green, gold, for profit, predatory); the decisions that are made in relation to it (in support of CC-BY licenses, for instance, but not article or author processing charges); the specific technologies, tactics, and strategies that are employed (the use of Open Journal Systems software to focus on the publication of open access journals rather than books, say); the particular historical, social, and legal conjunctions of time, situation, and context in which such decisions are made and practices, actions, and activities take place (in the humanities and social sciences [HSS] or STEM, Global North or South); and the networks, relationships, and flows of culture, community, society, and economics they encourage, mobilize, and make possible. In other words, the politics of openness access are the product of contingent, pragmatic decisions involving power, conflict, and violence, decisions that can nevertheless be disarticulated and transformed as a result of struggle, and perhaps even a new form of hegemony established: whether it be a particular flavor of open access or some other publishing system, based on peer-to-peer file sharing perhaps.[23] Open access is thus not necessarily a mode of left resistance.[24] Nevertheless, what is interesting about the transition to the open access publication and archiving of research is the way it is creating at least some openings that allow academics to destabilize and rethink scholarly publishing and, with it, the university, beyond the model espoused by free-market capitalism.

In fact, it could be argued that the open access movement currently possesses greater potential for doing so than a lot of supposedly more politically subversive initiatives. This is certainly the case with regard to the ability of open access to establish chains of equivalence between a range of different struggles, and thus garner a large constituency of supporters made up not just of academics and those associated with the free software and free culture movements, but of students, former students, and even representatives of capital itself. That said, for all its ability to create such openings, open access continues to operate for the most part within particular frames and limits. While John Willinsky presents it as "both a critical and practical step toward the unconditional university" imagined by Jacques Derrida in "The Future of the Profession or the University without Condition," the open access movement is actually quite conditional (at least currently). It may promote the "right to speak and to resist unconditionally" everything that concerns the restriction of access to knowledge, research, and thought, as Willinsky says. The open access movement does so for the most part, however, only on condition that the "right to say everything" about a whole host of other questions is not exercised.[25] Included in this are questions not just about the use of blogs, Facebook, Google+, and so on by open access advocates such as Suber and me, but also about the human, the sovereign, proprietorial subject, the individualized author, originality, the text, intellectual property, and copyright. But what if, taking our cue this time from Derrida, we were to view the open access movement as merely a strategic starting point for thinking about such issues? What if we were to regard the conditionality of open access not as a prompt to move beyond open access, or to leave it behind and replace it with something else, but rather as directing us to follow the logic of the open access movement through "to the end, without reserve," to the point of agreeing with it against itself?[26] What if we were to begin to speak about, and to resist unconditionally, some of the other orthodoxies that concern the restriction of access to knowledge, research, and thought: precisely ideas of human authorship and originality and the copyright system that sustains them? I single out copyright because if we wish to struggle against the becoming business of the university, then we have to accept that this may involve us in a struggle against the system of copyright too (rather than endeavoring to merely reform it), since the latter is one of the main ways in which knowledge, research, and thought are being commodified, privatized, and corporatized.[27]

Copyright

Drastically simplifying the situation for the sake of brevity, there are two key justifications for copyright in this context: that associated with economic rights and that connected with what is known as author's or moral rights, respectively.[28] In the former, which dominates the Anglo American copyright tradition, the emphasis is placed on the protection of the commercial interests of the author, producer, or distributor of a work and their right to benefit from it financially by making and selling copies. This is how the majority of conventional academic publishing firms regard the books they bring out: as commodities whose rights to their commercial exploitation have been transferred to them.

To be sure, few authors of research monographs derive substantial income directly from such writings. Most are willing to assign or license the rights to the commercial interests to publishers in return for having the resulting volumes edited, published, distributed, marketed, promoted, and, they hope, read and engaged with by others. In this respect, academics are operating on the basis that these activities have the potential to lead to further income indirectly: through a growth in their reputation and level of influence, and thus to greater opportunities for career advancement, promotion, salary increases, and so on. Consequently, it is publishers that are perceived as being most at risk financially from the infringement of copyright in this economic sense. Witness, with regard to Aaaaarg's "pirating" of texts drawn from theory, philosophy, politics, avant-garde fiction, and related areas (including some of my "own"), the fact that it was the self-professed "radical publishing house" Verso—and not the authors—who posted the December 2009 cease-and-desist letter asking the knowledge-sharing platform to take down copies of those titles by Žižek, Rancière, Badiou, and other authors for which Verso reserves the rights. They did so in spite of the fact that, as Janneke Adema points out, many people still prefer to read books especially in print form, thus "making the online and free availability of texts nothing more than a marketing tool for the sales of the printed version" (much as piracy of *Game of Thrones* has been "better than an Emmy" in helping to drive subscriptions to HBO, according to the chief executive of its parent company, Time Warner).[29] This marketing is particularly valuable for smaller, independent publishers and is presumably one reason more of them have not responded to Aaaaarg in a similar fashion, another being the negative publicity and backlash Verso's action provoked among the online community of readers of radical thought.[30]

Of course, some authors may wish to support independent publishers of radical political content. Many such presses are in a precarious financial situation, especially in comparison to their multinational conglomerate-owned rivals. They are heavily reliant on the income generated from the sale of books to which they own the rights to be able to stay in business and so bring out more such titles in the future. However, because the copyright system is one of the main ways in which knowledge, research, and thought are being commodified and privatized, it is perhaps more difficult for those committed to the struggle against the commercialization of culture and society to wholeheartedly support defenses against infringement on the basis of the protection of economic rights. After all, if we are interested in trying out new or different economic, legal, and political systems to that of capitalism (and not just neoliberalism), it can hardly come as a surprise if doing so should have implications for those publishing firms whose business models continue to depend on turning even such obviously political phenomena as Occupy, communism, and the revitalized student movement into commodities that can be privately owned and bought and sold.

Consider the recent dispute over the request of another independent radical publisher, Lawrence & Wishart, that the Marxists Internet Archive delete ten copies of the scholarly edition of the *Collected Works of Marx and Engels* for which the former owns copyright. Lawrence & Wishart—at one point in its history the Communist Party of Great Britain's publishing house—made this request because it wished to enter into an arrangement with a distributor to sell a digital version of the *Collected Works*, which runs to fifty volumes in all, to university libraries internationally, to be purchased out of public funds.[31] However, in the words of one volunteer at the Marxists Internet Archive, this has left Lawrence & Wishart in a situation where it "wants to spread the words of communism via a capitalistic method."[32]

When it comes to moral rights, the justification for copyright has its basis in the protection of what is held to be an inalienable right of authors in their work. This right, often positioned as originating in the culture of Western Europe and as operating in a supplementary, secondary, even marginal, relation to economic rights, applies to the work considered as an expression of the unique mind or personality of the author. It is this special connection, forged between author and work in the very act of creation, that is also perceived as bestowing the latter with its originality (rather than any sense of the work being novel or inventive). Consequently, in contrast to economic rights, the moral rights of the author cannot simply be waived, sold, or transferred to another individual or corporate entity such as a publisher.

Now some might argue that critical theory and radical philosophy's decentering of ideas of the subject and the human, and associated declaration of the "death of the author," has contributed to the expansion of the neoliberal globalized copyright industry and its shifting of the emphasis even further away from safeguarding the rights of the individual author as original creator and onto safeguarding the rights to a commodity that can be bought and sold for profit regardless of who created it. By the same token, if we are inclined to be generous, then the tendency on the part of many theorists to assert vigorously their authorship of particular works, ideas, and concepts, thus both enclosing and branding them by association with a proper name on the basis they are "all mine" (an original expression of their own unique selves), can be positioned as one attempt to make this shift in emphasis from culture and human authorship to economics and property ownership a little less smooth. From this latter perspective, the risk that copyright infringement poses to authors is more to their moral rights, and in particular (1) the right of attribution—which, to return to the example already employed, Aaaaarg does not tend to threaten, as the authors of most of the texts on the knowledge-sharing platform are clearly named and identified as such (you can browse its library by author surname); (2) the right of integrity, which enables authors to refuse to allow the original, fixed, and final form of a work to be modified or distorted by others—which, again, Aaaaarg tends not to be associated with infringing to the extent that many of the texts it features are scanned PDF copies of the original publications (although this right is infringed if and when Aaaaarg includes unauthorized translations, to provide just one example); and (3) the right of disclosure, which covers the right to determine who publishes the work, how, where, and in which contexts—and, to be sure, Aaaaarg may represent for some academics a loss of reputation, honor and esteem, to the degree that their work is being republished informally, outside the conventional institutional frameworks, and in places and ways other than those of their choosing.

The question we need to ask, though, is to what extent operating according to the moral rights of attribution, integrity, and disclosure leads theorists and philosophers to act to all intents and purposes as if they continue to subscribe to the idea of the author as individual creative genius that emerged from within the cultural tradition of European romanticism—a notion that the humanities' critical interrogation of the concepts of the subject, the human, and indeed the author is in many respects an attempt to challenge. It is precisely this romantic belief that underpins the idea of the work as the original expression of the unique personality or consciousness of the human author and on which such moral rights are in turn based.

This is not to imply we should necessarily do away with the concept of the author. Yet what this argument does suggest, at the very least, is that we need to explore further how radical thought can enact ideas of authorship in ways that do not either slip back into compulsively repeating a version of romantic individualism and its notions of originality, or empty this out so that texts merely become exchangeable commodities. To provide one example of how we might begin to do so, could we try acting something like pirate philosophers?

Pirate Philosophers

Of course, and as Adrian Johns makes clear, despite its romantic, countercultural image, much of the philosophy associated with online piracy today is itself a "moral philosophy through and through." It is concerned "centrally with convictions about freedom, rights, duties, obligations, and the like." What is more, it is a philosophy that has its historical roots in a "marked libertarian ideology": one of the UK pirate radio ships of the 1960s was even called the *Laissez Faire*. From a British perspective, it is a philosophy that "helped to make Thatcherism in particular what it was," Johns notes.[33] So piracy in this sense is not opposed to capitalism; it is fundamental to it.

Such pirate philosophers as we might envisage here would have to try acting like pirates more in the classical sense of the term. When the word *pirate* begin to appear in the texts of the ancient Greeks (as chapters 1 and 4 show), it was directly related to the verb *piraō*, to "make an attempt, try, test, get experience, endeavor, attack."[34] It is here that the modern expression *pirate* (which in Greek is also connected to "tease," to "give trouble") has its etymological origins. This was long before the seventeenth century, when Johns positions it as being routine to use the term in England, or circa 1710, when he has its use when referring to acts of intellectual misconduct becoming an everyday one.[35]

In this respect, what is most interesting about certain phenomena associated with networked digital culture such as Napster, the Pirate Bay, Popcorn Time, Aaaaarg, and the unauthorized downloading from JSTOR's database of the open access guerrilla Aaron Swartz,[36] is that we cannot tell at the time of their initial appearance whether they are legitimate. This is because the new conditions created by networked digital culture, for example the ability to digitize and make freely available whole libraries worth of books (as is the case with Google Books and Aaaaarg), at times require the creation of equally new intellectual property laws and copyright policies. The 1998 Digital Millennium Copyright Act, 2001 European Union Copyright Directive, and the UK's Digital Economies Act 2010 are some examples; the Google Book settlement, SOPA (Stop Online Piracy Act), and PIPA (Protect IP Act) in the United States are (or were) others. It follows that we can never be sure whether these so-called pirates, in the attempts they are making to contend with the new conditions and possibilities created by networked digital culture, to try them and put them to the test, are not in fact involved in the creation of the very new laws, policies, clauses, settlements, licensing agreements, and acts of Congress and Parliament by which they could be judged.

Consider the case of William Fox, a filmmaker who relocated from America's East Coast to California in the early twentieth century in part "to escape controls that patents granted the inventor of filmmaking, Thomas Edison." As Lawrence Lessig recounts in his chapter on "pirates" in *Free Culture*, Fox founded the film studio 20th Century Fox precisely by pirating Edison's creative property.[37] (Ironically, the chairman and CEO of 20th

Century Fox's owners, 21st Century Fox, is that scourge of Internet piracy Rupert Murdoch, who attacked the Obama administration on Twitter after the White House indicated it would not be supporting some of the harsher measures proposed in the SOPA bill.)[38] As the example of Fox shows, we can never tell the founder of a new institution or culture in advance. We can only finally judge whether the activities of such supposed pirates are legal or not, legitimate or not, just or not from some point "projected into an indefinite future."[39] This is why I am suggesting perhaps acting something like pirate philosophers: because a responsible ethical, as opposed to moralistic, approach to piracy would not presume to know what it is in advance.[40]

Another way to think about the issue of piracy is in relation to the legislator in Rousseau's *The Social Contract*. Here, too, we can never know whether the legislator—the founder of a new law or institution, such as a university, or indeed new way of being and doing as a theorist or philosopher—is legitimate or a charlatan. The reason for this is the aporia that lies at the heart of authority, whereby the legislator already has to possess the authority the founding of the new institution is supposed to provide him or her with in order to be able to found it. Certain so-called Internet pirates are in a similar situation to Rousseau's legislator. They too may be involved in performatively inventing, trialing, and testing the very new laws and institutions by which their activities may then be judged and justified. As such, they can claim legitimacy only from themselves. This is a state of affairs that as well as marking their impossibility also constitutes their founding power, their instituting force. It is here, between the possible and the impossible, legality and illegality, that we must begin any assessment or judgment of them. And it should be noted that it is not just the potential pirates who may be legislators or charlatans. The current laws and institutions by which we might condemn Internet piracy as illegal are based on the same aporetic structure of authority. Such lawmakers are always also undecidably charlatans or pirates too (or hackers, in the case of Murdoch's News International—now News UK).[41]

Consequently, we cannot tell what will happen with pirate philosophy. It may lead to new forms of culture, economy, and education more in tune with the change in political mood post-2008: where people work and create for reasons other than to get paid; where the protection of copyright is no longer possible; where the cultural industries—book publishers, the press, and so forth—are radically reconfigured; music, television, and film are available to freely stream, download, and share (which is already the case); academic monographs are circulated using text-sharing platforms (which they already are); and even our concepts of the unified, sovereign, proprietorial subject and individualized humanist author (on which, as chapter 1 shows, the Creative Commons, open access, and free software movements all depend) are dramatically transformed. In this respect, pirate philosophy may play a part in the development of not just a new kind of university, but new laws, new economies, and new ways of organizing postindustrial society. In the process, it may have as profound an effect "as the establishment of copyright ... in the eighteenth century, and the development of modern patent systems in the nineteenth," to borrow Johns's words.[42] But it may not. And that is the point. As with the famous remark about the significance of the French Revolution—let alone the "crisis of capitalism" and the "global springs"—it is still too early to tell.[43] Nevertheless, what is interesting is the potential that pirate philosophy contains for the development of a new kind of economy and society: one based far less on individualism, possession, acquisition, accumulation, competition, celebrity, and ideas of knowledge, research, and thought as something to be owned, commodified, communicated, disseminated, and exchanged as the property of single, indivisible authors (who, as Andrew Ross notes, are often likely to be corporate entities).[44]

Without a doubt, many currently at work in the university are going to experience any such trying or testing of the idea of acting something like pirate philosophers as an attack not just on copyright and the corporatization and marketization of the university, but on their professional identities too: as a challenge to the secure ground on which they have been operating for so long, based as that is on quite orthodox concepts of authorship, originality, and so forth. And their fears will be justified. Yet in order to respond to the forces of late capitalist society, do we not have to take the risk of leaving the safe harbor of our profession as it currently stands? After all, it is not as if we are going to be secure if we do nothing; our professional identities are already under threat. Might embarking on such an endeavor offer us a means of contending affirmatively with some of the forces behind this threat without simply succumbing to them, reacting with nostalgia or romanticism to them, or naively celebrating and assisting them?

Chapter 6 expands the principles developed in chapters 1 to 5. It offers further examples of what it might mean to act as something like pirate philosophers in the sense of performatively trialing new ways of composing, publishing, and circulating knowledge and research. But it also shows how certain elements of these new ways of being, thinking, and doing are already present within the academy (e.g., the publication of work in multiple places, including in draft, preprint, and gray literature form, with little attention often being paid to copyright agreements). In doing so, chapter 6 demonstrates how my argument here regarding the importance of not only what a theorist writes, but also the theory he or she acts out and performs, relates to my own publication with a legacy press of a print-on-paper codex book that takes pirate philosophy as one of its subjects.

6 THE FUTURE OF THE BOOK

The Unbound Book

It was well before computers that I risked the most refractory texts in relation to the norms of linear writing. It would be easier for me now to do this work of dislocation or typographical invention—of graftings, insertions, cuttings, and pastings—but I'm not very interested in that any more from that point of view and in that form. That was theorized and that was done—then. The path was broken experimentally for these new typographies long ago, and today it has become ordinary. So we must invent other "disorders," ones that are more discreet, less self-congratulatory and exhibitionist, and this time contemporary with the computer.[1]—Jacques Derrida

It is often said that the book today is being dramatically disrupted—that in the era of online authorship, comment sections, personal blogs, embeddable videos, apps, and texts being generally connected to a network of other information, data, and mobile media environments, the book is in the process of being diluted, dislocated, dispersed, displaced. If the book is to have any future at all in the context of these other modes of reading, writing, and forms of material occurrence, it will be in unbound form—a form that, while radically transforming the book may yet serve to save it and keep it alive.[2] Yet what is the unbound book? Can the book be unbound?

The *Oxford Dictionary Online* defines the term *bound* as follows:

bound *in* bind ... tie or fasten (something) tightly together ... ;
... walk or run with leaping strides ... ;
... a territorial limit; a boundary ... ;
... going or ready to go towards a specified place ... ;
past and past participle of bind.[3]

In this case, the unbound book would be one that *had* been gathered together and firmly secured, as a pile of pages can be to form a print-on-paper codex volume; *had* a certain destiny or destination or had been prepared, going or ready to go toward a specific place (as in "homeward bound"), such as perhaps an intended addressee, known reader, or identifiable and controllable audience; and *had* been springing forward or progressing toward that place or destiny in leaps and bounds. *Had,* because the use of the past participle suggests such binding is history as far as the book is concerned—that after centuries of print such conventional notions of the book have become outdated.

As we know from Ulises Carrión, however, there is no such thing as an unbound book. "A writer, contrary to the popular opinion, does not write books," he declares in "The New Art of Making Books" of 1975:

> A writer writes texts.
>
> The fact, that a text is contained in a book, comes only from the dimensions of such a text; or, in the case of a series of short texts (poems, for instance), from their number.[4]

The book is just a container for text. The idea of binding is thus essential to the book.

Tempting though it may be, then, we cannot say that whereas in the past the book *had* been bound, it no longer is; it has become unglued, unstuck. One reason we cannot say this is because e-book readers and iPad apps, while offering different types of binding to printed books, different ways of securing pages together, nevertheless reinforce rather conservative, papercentric notions of bookishness that are not only designed to mimic the traditional reading experience, but also make their identities every bit as closed, fixed, stable, and certain in their own ways as those of the scroll and codex (for authors and publishers, but also for readers). The main reason we cannot say this, however, is that an unbound book is quite simply no longer a book. Without a binding, without somehow being tied, fastened, or stuck together, a writer's text is not a book at all: it is just a text or collection of texts. *A text is a book only when it is bound.*[5]

Carrion's primary concern is with the conception of the book as an object—which he sees as a series of pages both divided and gathered together in a coherent, and usually numbered, sequence—and with its material forms of occurrence and fabrication: its parchment, cloth, wood pulp–based paper, ink, printing, typography, design, layout, and so forth. Is it possible, therefore, that rather than in ontological terms, the idea of the unbound book can be addressed more productively from the perspective of one of the other ways in which books can be said to be tied? I am thinking in terms of legal contracts. These function to establish territorial boundaries marking when certain ideas and actions relating to the book are "out of bounds," forbidden, limited by restrictions and regulations concerning copyright, intellectual property, notions of authorship, originality, attribution, integrity, disclosure, and so on.[6]

McKenzie Wark's 2007 article, "Copyright, Copyleft, Copygift," offers an interesting starting point for thinking further about the legally bound aspect of the book. In it Wark addresses the contradiction involved in his having on the one hand written a book against the idea of intellectual property, *A Hacker Manifesto*, and on the other hand published it with an established academic print press, Harvard University Press (HUP), which refused to allow him to release a copy under a Creative Commons license as part of the then still relatively new and emergent digital gift economy.[7] Wark's solution in "Copyright, Copyleft, Copygift" is to "live the contradictions!" between commodity and gift culture and also to carry a memory stick to speaking events so anyone who wants a postprint copy of *A Hacker Manifesto* can get one for free from him personally, in the form of a text file they can even alter if they so wish.[8] Nevertheless, disseminating *A Hacker Manifesto* by sneakernet—or pink Roos, in Wark's case—does little to resolve the problem he identifies: how to meet an author's desire to have his or her work distributed to, respected, and read by as many people as possible—something a legacy print press like HUP can deliver—while also being part of the academic gift economy.[9] Quite simply, books made available on a free offline access basis circulate much more slowly and far less widely than those made available for free online.[10] They also tend to carry less authority.

Wark does not appear to be aware in "Copyright, Copyleft, Copygift" of the possibility of self-archiving his research open access, thus making a copy of it available online for free to anyone with access to the Internet, without the need on the part of readers to pay a cover price, library subscription charge, or publisher's fee. Yet even if he were, open access self-archiving—green open access—does not provide a straightforward solution to Wark's dilemma, since there is an important difference between publishing scholarly journal articles open access and publishing books open access. As is made clear in the *Self-Archiving FAQ* written for the Budapest Open Access Initiative:

> Where exclusive copyright has been assigned by the author to a journal publisher for a peer reviewed draft, copy-edited and accepted for publication by that journal, then *that draft* may not be self-archived [on the author's own website or in a central, subject, or institutional repository] by the author (without the publisher's permission).

> The pre-refereeing preprint, however, [may have] already been (legally) self-archived. (No copyright transfer agreement existed at that time, for that draft.)[11]

This is how open access self-archiving is able to elude many of the problems associated with copyright or licensing restrictions with regard to articles in peer-reviewed journals (assuming the journals in question are not themselves already online and open access). However, "where exclusive copyright ... has been transferred ... to a publisher"—for example, "where the author has been paid ... in exchange for the text," as is generally the case in book publishing but not with journal articles—it may be that authors are not legally allowed to self-archive a copy of their book or any future editions derived from it open access at all. This is because, although the "text is still the author's 'intellectual property' ... the exclusive right to sell or give away copies of it has been transferred to the publisher."[12]

So what options are available to book authors if, like Wark, they wish to have their work read beyond a certain "underground" level (in Wark's case, that associated with net art and net theory), while at the same time being part of the academic gift economy?[13]

1. Authors can publish with an open access press such as Australian National University's ANU E Press, Athabasca University's AU Press, MayFly Books, or Open Book Publishers.[14] Graham Harman has brought out *Prince of Networks: Bruno Latour and Metaphysics* with re.press (see chapter 1), and John Carlos Rowe has published *The Cultural Politics of the New American Studies* with Open Humanities Press, both open access presses.[15] Still, with the best will in the world, few open access book publishers are established and prestigious enough as yet to have the kind of brand-name equivalence to HUP that Wark desires—especially when it comes to impressing prospective employers and getting work reviewed and generally accepted by the gatekeepers of intellectual property (the university, the library, the research funder, and so on)—although it is, we can hope, only a matter of time.
2. Authors can insist on only signing a contract with a press that will allow them to self-archive a peer-reviewed and perhaps even copyedited version of their book. The difficulty, of course, is finding a brand-name publisher willing to agree to this.
3. Authors can sign a contract with a press that is not usually amenable to open access and then endeavor to negotiate—as Wark did (unsuccessfully) with HUP—to see if it would be willing to make the published version of their book available for free online, with only the printed version available

for sale. Authors who have published in this way include Ted Striphas with *The Late Age of Print* from Columbia University Press and Gabriella Coleman with her book *Coding Freedom: The Ethics and Aesthetics of Hacking* from Princeton University Press.[16] However, such instances seem to be regarded by many publishers as little more than occasional experiments—the dipping of a toe in the open access water, as it were, in order to test the market.[17]

4. Authors can adopt a variation of the strategy advocated on the *Self-Archiving FAQ* written for the Budapest Open Access Initiative with regard to scholarly journal articles. This is simply "don't-ask/don't-tell." Instead, authors can publish with whatever publisher they wish, self-archive the full text (as many are now doing on the Academia.edu social networking platform, if not in open access repositories), "and wait to see whether the publisher ever requests removal."[18]
5. Authors can wait for someone to publish a *pirate* copy of their book on a text-sharing network such as Aaaaarg or LibGen (indeed, some are already doing so themselves).

Noticeably, all these strategies in effect fasten what are identified—conceptually, economically, temporally, materially, and morally—as finished, complete, unified, and bound books in legal binds; they are just different ways of negotiating such binds. What, though, if book authors were to pursue ways of openly publishing their research before it is tied up quite so tightly?

To investigate this idea, in June 2010 I began experimenting with an Open Humanities Notebook, taking as one model for doing so the Open Notebook Science of the organic chemist Jean-Claude Bradley.[19] As is emphasized in a September 2010 interview with the independent journalist and open access advocate Richard Poynder, Bradley is making the "details of every experiment done in his lab"— the whole research process, not just the findings—freely available to the public on the Web. This "includes all the data generated from these experiments too, even the failed experiments." What is more, he is doing so in "real time," "within hours of production, not after the months or years involved in peer review."[20]

Since one of my books-in-progress deals with a series of projects that use digital media to actualize, or creatively perform, critical theory and philosophy, it seems appropriate to make the research for this volume freely available online in my Open Humanities Notebook—and to do so on a basis that permits it to be distributed, reproduced, transmitted, translated, modified, remixed, built on, used, and "pirated" in any medium, even without indication of "origin."[21] I am making my research available in this way more or less as it emerges, not just in draft and preprint form as journal articles, book chapters, exhibition catalog essays, and so on, but also as contributions to e-mail discussions, conference papers, and lectures—long before any of these texts are collected together and given to a publisher to be bound as a book (although the process of making the research related to this project freely available online can of course continue afterward too, after print or e-publication).[22]

As is the case with Bradley's Open Notebook Science, this Open Humanities Notebook acts as a space where the research for my book-in-progress, provisionally titled *Media Gifts*, can be disseminated quickly and easily in a manner that enables it to be openly shared and discussed. (Working in this fashion can thus be considered not just as a form of open writing to set alongside that of Kathleen Fitzpatrick, say [see chapters 1 and 5], but of open humanities research.) Yet more than that, this open notebook provides me with an opportunity to experiment critically with keeping at least some of the ties that are used to bind books once a text has been contracted by a professional press relatively loose.

For instance, it is common for most book contracts to allow authors to retain the right to reuse material that has previously appeared elsewhere (e.g., as scholarly articles in peer-reviewed journals), in their own written or edited publications, provided the necessary permissions have been granted. What, though, if draft or preprint versions of the chapters that make up my book are initially gathered together in this open notebook, albeit perhaps in a nonlinear fashion? When it comes to eventually publishing this research as a bound book, are brand-name legacy presses likely to reject it on the grounds of potential reduced sales since a version of this material will already be available online?[23]

One possibility is that I will be required to remove any draft or preprint versions of these chapters from my Open Humanities Notebook to ensure the publisher has the exclusive right to sell or give away copies. This is what happened to Ted Striphas with regard to an article he wrote, "Performing Scholarly Communication," that was published in the January 2012 issue of the Taylor & Francis journal *Text and Performance Quarterly*. Taylor & Francis's publication embargo stipulated that Striphas could not make the piece available on a public website, in any form, for eighteen months from the date of publication. So Striphas had to take down the preprint version that was available on his Differences and Repetitions wiki, a site where he publishes drafts of his writings-in-progress on what he terms an "open-source" and "partially open source" basis.[24]

Another possibility is that making at least some draft or preprint versions of this research available in my Open Humanities Notebook will be seen by the press as a form of valuable advance exposure, marketing, and promotion. If so, the question then will concern how much of the book can be gathered together in this fashion before it becomes an issue for the publisher. At what point does the material that goes to make up a book become bound tightly enough for it to be understood as actually making up a book? Where in practice is the line going to be drawn?

What if some of this work is disseminated out of sequence, under different titles, in other versions, forms, times, names, and places where it is not quite so easy to bind, legally, economically, temporally, or conceptually, as a book? Let me take as an example the chapter in *Media Gifts* that explores the idea of Liquid Books.[25] A version of this material appears as part of an actual "liquid book" that is published using a wiki and is thus free for users to read, comment on, rewrite, remix, and reinvent.[26] Meanwhile, another "gift" in the series, a text on pirate philosophy, is currently available only from "pirate" peer-to-peer networks. There is no original or master copy of this text in the conventional sense. "Pirate Philosophy" exists only to the extent it is part of "pirate networks" and is "pirated."[27]

Indeed, while each of the media gifts the book is concerned with—at the time of this writing, there are more than ten—constitutes a distinct project in its own right, they can also be seen as forming an extended network or meshwork of dynamic relations that pass between a number of different texts, websites, archives, wikis, Internet TV programs, and other online and offline traces.[28] Consequently, if *Media Gifts* is to be thought of as a book, it should be understood as an open, decentered, distributed, nonlinear (in both form and content), multilocation, multimedium, multiple-identity book.[29] While a version maybe indeed appear at some point in print-on-paper or online book form, some parts and versions of it are also to be found on blogs, others on wikis, others again on video-sharing platforms and file-sharing networks.[30] To adapt a phrase of Maurice Blanchot's from *The Book to Come* (for whom Stéphane Mallarmé's "*Un coup de dés* orients the future of the book both in the direction of the greatest dispersion and in the direction of a tension capable of *gathering* infinite diversity, by the discovery of more complex structures"), *Media Gifts* is a book "gathered through dispersion."[31]

That said, we do not need to go quite this far in dispersing our books if we just want to establish a publishing strategy others can adopt and follow.

Prior to publication, Wark had already disseminated versions of *The Hacker Manifesto* on the Internet as work-in-progress by means of the nettime mailing list especially. It is an authorial practice that is increasingly common today, down to the level of e-mails and social media and social networking posts; and most presses are willing to republish material that has previously appeared in these forms. Still, what if authors provide interested readers with something as simple as a set of guidelines and links showing how such constellations of texts can be bound together in a coherent, sequential form (perhaps using a collection and organization tool such as Anthologize, a plug-in that uses WordPress to turn distributed online content into an electronic book)?[32] Just how dispersed, loosely gathered and structured does a free, open, online version of a book have to be, spatially and temporally, for brand-name presses to be prepared to publish a bound version?

In the essay "The Book to Come," Jacques Derrida asks: "What then do we have the right to call 'book' and in what way is the question of *right*, far from being preliminary or accessory, here lodged at the very heart of the question of the book? This question is governed by the question of right, not only in its particular juridical form, but also in its semantic, political, social, and economic form—in short, in its total form."[33]

My question is: What do we have the right *not* to call a "book"?

Dispersing our current work-in-progress will not only provide us with a way of keeping some of the legal ties that bind books loose but may also help us to think differently about the idea of the book itself. Graham Harman writes with regard to philosophy:

> In not too many years we will have reached the point where literally anyone can publish a philosophy book in electronic form in a matter of minutes, even without the least trace of official academic credentials. I don't bemoan this at all—the great era of 17th century philosophy was dominated by non-professors, and the same thing could easily happen again. As far as publishing is concerned, what it means is that all publishing is destined to become vanity publishing. (Alberto Toscano recently pointed this out to me.) You'll just post a homemade book on line, and maybe people will download it and read it, and maybe you'll pick up some influence.[34]

Yet what is so interesting about recent developments in digital publishing is not that, what with open access, WordPress, Scrib'd, Booki, and Aaaaarg, producing and distributing (and even selling in the case of Smashwords and Kindle Direct Publishing) a book is something nearly everyone can do today in a matter of minutes. It is not even that book publishing may, as a result, be becoming steadily more like blogging or vanity publication, with authority and certification provided as much by an author's reputation or readership, or the number of times a text is visited, downloaded, cited, referenced, linked to, blogged about, tagged, bookmarked, ranked, rated, or "liked," as it is by conventional peer review or the prestige of the press. All of these criteria still rest on and retain fairly conventional notions of the book, the author, publication, and so forth. What seems much more interesting is the way certain developments in digital publishing contain at least the potential for us to regard the book as something that is not fixed, stable, and unified, with definite limits and clear material edges, but as liquid and living (see chapter 4), open to being continually and collaboratively produced, written, read, edited, annotated, critiqued, updated, shared, supplemented, revised, reordered, reiterated, and reimagined. Here, what we think of as *publication*—whether it occurs in real time or after a long period of reflection and editorial review, all at once or in fits and starts, in print-on-paper or electronic form—is no longer an end point. Rather publication is just a stage in an ongoing process of spatial and temporal unfolding.

What I have been describing in terms of work-in-progress is very much part of a new strategy for academic writing and publishing that I and more than a few others are critically and creatively experimenting with at the moment. One of the aims of this strategy is to move away from thinking of open access primarily in terms of scholarly journals, books, and even central, subject, and institutionally based self-archiving repositories. Instead, the focus is on developing a (pre- and post-) publishing economy characterized by a multiplicity of different, and at times conflicting, models and modes of creating, binding, collecting, archiving, storing, searching, reading, and interacting with academic research and publications.

This new publishing strategy has its basis in a number of speculative gambles with the future. It challenges some long-held assumptions by suggesting, among other things (and whether rightly or wrongly):

• That the correct, proper, and most effective forms for creating, publishing, disseminating, and archiving academic research will be progressively difficult to determine and control. Scholars will continue to write and publish paper and papercentric texts. However, more and more they will also generate, distribute, and circulate their research in forms that are specific to image and Internet-based media cultures, and make use of video, film, sound, music, photography, data, graphics, animation, augmented reality, 3D technology, geolocation search capabilities, and hybrid combinations thereof. (The Article of the Future project from the academic publisher Elsevier is already pointing in this direction, as are PLoS Hubs.)[35]

• That scholars will be far less likely to publish a piece of academic research in just one place, such as a tightly bound book or edition of a peer-reviewed journal produced by a brand-name press.[36] Again, they will no doubt still place their work in such venues. Nevertheless, their publishing strategies will be far more pluralistic, decentered, distributed, multifaceted, and liquid than they have been to date, with researchers—motivated in some cases to be sure by a desire to increase the size of their academic footprint—making simultaneous use of WordPress, MediaWiki, Monoskop, YouTube, Vine, iTunesU, Meerkat, and whatever their future equivalents are to disseminate and circulate their research in a wide variety of places, contexts, voices, and registers. (Moreover, they will often do so without showing too much concern for copyright, as the small amount of money at stake means most such agreements will not be enforced.) It is even possible that with the further development of open access, open data, and open education resources, we will move to a situation where the same material is reiterated as part of a number of different, interoperable texts and groupings or, as Derrida speculates, where research is no longer grouped according to the "corpus or opus—not finite and separable oeuvres; groupings no longer forming texts, even, but open textual processes offered on boundless national and international networks, for the active or interactive intervention of readers turned coauthors, and so on."[37]

• That an increasing number of scholars will create, publish, and circulate their written research not just as long- or even medium-length forms of shared attention along the lines of Amazon's Kindle Singles, Ted Books (part of the Kindle Singles imprint), Palgrave Pivot, and Stanford Literary Lab pamphlets,[38] but in modular or chunked forms too: from the "'middle state' (between a blog and a journal") posts of "The New Everyday" section of Media Commons,[39] down to the level of passages, paragraphs, and at times even perhaps sentences (i.e., nanopublishing).[40] Scholars will do so to facilitate the flow of their research and the associated data and metadata between different platforms and other means of support: books, journals, websites, and archives, but also e-mails, blogs, tweets, wikis, discussion forums, chatstreams, podcasts, text messages, e-book readers, and tablets. These are places where, depending on the particular system and platform, it can be commented and reflected on, augmented, mined, visualized, hacked, remixed, reflowed, reversioned, and repurposed. As Johanna Drucker notes with regard to how these new, often microformats and genres may be accounted for within the metrics of academic communities when it comes to ranking a scholar's achievement at moments of promotion or tenure, "the possibilities are rapidly becoming probabilities with every sign that we will soon be tracking the memes and tropes of individual authors through some combination of attribute tags, link-back trails, and other identifiers that can generate quantitative data and map a scholar's active life."[41]

- That scholars will also publish, disseminate, and circulate their research in beta, preprint, and gray literature form (as the Public Library of Science is already doing to a limited extent with *PLoS Currents: Influenza*, as is PressForward).[42] In other words, they will publish and archive the pieces of paper, e-mails, websites, blogs, or social networking posts on which the idea is first recorded, and any drafts, working papers, or reports that are circulated to garner comments from peers and interested parties, as well as the finished, peer-reviewed, and copyedited texts.
- That many scholars and academic journals will publish the data generated in the course of research, with a view to making this source material openly and rapidly available for others to forage through, shape, and bind into an interpretation, narrative, thesis, article, or book (see both figshare and the *Journal of Open Archaeology Data* for examples).[43]
- That much of the emphasis in institutional publishing, archiving, and dissemination strategies will switch from primarily capturing, selecting, gathering together, and preserving the research and data produced by scholars and making them openly accessible, to placing more emphasis on actively and creatively doing things with the research and data that are being continually selected, gathered, and made accessible. This will be achieved not least by both institutions and scholars offering users new ways to acquire, read, write, manage, interpret, and engage with their research, references, and data, both individually and collaboratively, pre- and postpublication; and in the process create new texts, objects, activities, and performances from this source material (as in the case of such early adopters as CampusROAR at the University of Southampton or the Larkin Press with its aim to provide "a web interface for authors and editors to create, manage and disseminate multi-format academic output [eBook and Print] from The University of Hull, combining existing University activities into a publishing whole").[44] It is even conceivable that the process of creating new texts, objects, activities, and performances

from this source material—including bringing groups of people together; organizing, educating, training, and supporting them; providing the appropriate platforms, applications, and tools; and so on—will become the main driver of research, with the production of papercentric texts such as books and journal articles being merely a by-product of this process rather than one of its end goals.

Since we are thinking about decentered and multiple publishing networks, a question needs to be raised at this point concerning the agency of both publishers and authors: Who is it that is experimenting with this new economy exactly?

I am aware of writing "I" a lot here—as if, despite everything, I am still operating according to the model whereby the work of a theorist or philosopher such as myself is regarded as being conceived, created, and indeed signed by a static and stable, individualized human author, and presented for the attention of a reader who, even for Derrida, can "interrogate, contradict, attack, or simply deconstruct" its logic, but who "cannot and must not change it."[45] In fact, the series of projects I have been referring to regarding my work-in-progress not only involve me in employing numerous and at times conflicting figures, voices, registers, and semiotic functions (multiple differential "I"s, as it were); they also arise out of what can be thought of as the processual and relational interconnections between a multiplicity of different authors, groups, institutions, and actors. This multitude includes those operating under the names of Culture Machine, Open Humanities Press, the Open Media Group, and Centre for Disruptive Media.[46]

Mark Amerika must be included in this list too, as an earlier version of this chapter was written as a contribution to his remixthebook project.[47] It is a remix of his "Sentences on Remixology 1.0," which is itself a remix of Sol LeWitt's "Sentences on Conceptual Art."[48] So when I say "I" here, it means at least all of the above. Yet it means even more than that, since some of the projects I am involved with and which feature in *Media Gifts* are also open to being collaboratively and even anonymously written.[49] Remixing Amerika remixing Alfred North Whitehead, it is what might be thought of as stimulating "the production of novel togetherness"—a togetherness made up of neither singularities, nor pluralities, nor collectivities.[50] In this sense it is not possible to say exactly who, *or what*, "I" am or "we" are.

"What does it mean to go out of oneself?" Am I unbound? Out of bounds? Is all this unbound?

Channeling Mark Amerika again, we should think of any writer or theorist such as myself as a medium, sampling from the vocabulary of critical thought. In fact if you pay close attention to what I am doing in this performance, you will see I am mutating myself—this pseudo-autobiographical self I am performatively constructing here—into a kind of postproduction processual medium. Just think of *me* as a postproduction of presence.

This chapter began by suggesting the word *book* should not be applied to a text generated in such a way, as without being tied or fastened tightly together—by the concept of an identifiable human author, for example—such a text is not a book at all: it is "only" a text or collection of texts.

To sample Sol LeWitt, we could say that one usually understands the texts of the present by applying the conventions of the past, thus misunderstanding the texts of the present. That, indeed, is one of the problems with a word such as *book*. When it is used—even in the form of *e-book, unbound book, unbook,*[51] or *the book to come*—it connotes a whole tradition and implies a consequent acceptance of that tradition, thus placing limitations on the writer or theorist who would be reluctant to create anything that goes beyond it.

Then again *book* is perhaps as good a name as any other since books historically have always been more or less loosely bound. For example, the Codex Sinaiticus, created around 350 AD, is one of the two oldest surviving Bibles in the world (the other is the Codex Vaticanus in Rome). As it currently exists, the Codex Sinaiticus, which contains the earliest surviving copy of the Christian New Testament and is the antecedent of all modern Christian Bibles, is incomplete. Nevertheless, it still includes the complete New Testament, half of the Old Testament, and two early Christian texts not featured in modern Bibles, all gathered into a single unit. So it is one of the first Bibles as we understand it. Yet more than that, it is arguably the first large bound book, as to gather together so many texts that had previously existed only as scrolled documents required a fundamental advancement in binding technology, a process that resulted eventually in the scroll or roll giving way to the codex book.[52]

Just as interesting is the fact that the Codex is also the most altered early biblical manuscript, containing approximately 30 corrections per page, roughly 23,000 in all. And these are not just minor corrections. At the beginning of Mark's Gospel, Jesus is not described as being the son of God. This was a revision. In the Codex Sinaiticus version, Jesus becomes divine only after he is baptized by John the Baptist. Nor is Jesus resurrected in the Codex Sinaiticus. Mark's Gospel ends with the discovery of the empty tomb. The resurrection takes place only in competing versions of the story found in other manuscripts.[53] Nor does the Codex contain the stoning of the adulterous woman, "Let he who is without sin cast the first stone," or Jesus's words on the cross, "Father forgive them for they know not what they do."

So the Bible—often dubbed "the Book of Books"—cannot be read as that most fixed, standard, permanent, and reliable of texts, the unaltered word of God. On the contrary, when the Codex Sinaiticus was created in the middle of the fourth century, the text of the Bible was already seen as being collaborative, multiauthored, fluid, evolving, emergent.

Shakespeare's First Folio provides another example from this history of fluid, emergent books.[54] As Adrian Johns shows, this volume includes "some six hundred different typefaces, along with nonuniform spelling and punctuation, erratic divisions and arrangement, mispaging, and irregular proofing. No two copies were identical. It is impossible to decide even that one is 'typical.'" In fact, according to Johns, it is not until 1790 that the first book regarded as having no mistakes was published.[55]

We could therefore say that books have always been liquid and living to some extent.[56] Digital media technologies and the Internet have simply helped to make us more aware of the fact.[57]

Indeed, if I am interested in the domains of electronic books and publishing at all, it is because the defamiliarization effect produced by the change in material occurrence from print-on-paper to those associated with networked digital media technologies offers us a chance to raise the kind of questions regarding our ideas of the book (but also of the unified, sovereign, proprietorial subject; the individualized author, the signature, the proper name; originality, fixity, the finished object; the canon, the discipline, tradition, intellectual property; the Commons, community, and so on), we should have been raising all along. As I have endeavored to show at length elsewhere, such questions were already present with regard to print and other media. However, as a result of modernity and the "development and spread of the concept of the author, along with mass printing techniques, uniform multiple-copy editions, copyright, established publishing houses, editors," and so forth, they have "tended to be taken for granted, overlooked, marginalized, excluded, or otherwise repressed."[58] Consequently, books have taken on the impression of being much more fixed, stable, static, reliable, permanent, authoritative, standardized, and tightly bound than they actually are or have ever been. For even if a book is mass-produced in a multiple-copy print edition, each copy is different, having its own singular life, agency, history, old age, and death, which is why we can form affective and symbolic attachments to them.

This is not to say we have never been modern, that books have never been tightly fastened or bound, just that this force of binding is what modernity, and the book, is ... or, perhaps, was.

Notes

Preface

1. Alain Badiou, *The Rebirth of History: Times of Riots and Uprisings* (London: Verso, 2012), 5.

2. Maurice Merleau-Ponty, "Sartre, Merleau-Ponty: Les Lettres d'une rupture," *Magazine Littéraire*, no. 320 (April 1994), cited in Wendy Brown, *Politics Out of History* (Princeton, NJ: Princeton University Press, 2001), 43. "In terms of" is how Michel Foucault puts it when he is asked if he wrote *The Use of Pleasure* and *Care of the Self* "for the liberation movement": Michel Foucault, "The Concern for Truth," in *Politics, Philosophy, Culture: Interviews and Other Writings, 1977–1884*, ed. Lawrence D. Kritzman (New York: Routledge, 1988), 263.

Chapter 1

1. "Etymology of Pirate," *English Words of (Unexpected) Greek Origin*, http://ewonago.wordpress.com/2009/02/18/etymology-of-pirate/, posted by Johannes, February 18, 2009, accessed May 22, 2015.

2. For one much discussed popular expression of the postcrash rejection of political representation, see the comedian Russell Brand's critique of mainstream politics in the UK. Originally broadcast October 23, 2013, on the BBCs *Newsnight* television program, a clip of the interview with Brand conducted by presenter Jeremy Paxman has now been viewed over 11 million times on YouTube and was one of the ten most watched in the UK (http://www.youtube.com/watch?v=3YR4CseY9pk). Further variations on this theme are provided by the "horizontal" grassroots groundswells against the "old politics" of the media and political establishment that were such notable features of the 2014 Scottish independence referendum and the election of Jeremy Corbyn as leader of UK Labour Party in 2015.

It is also worth emphasizing that while the post-2008 political protests may constitute something new in terms of their scale and intensity, in many countries they are actually the product of long periods of struggle. As Mikkel Bolt Rasmussen points out, for example, "Egypt experienced the most extensive wave of strikes ever seen in the country" between 2006 and 2008, while the Gafsa region in Tunisia suffered a "kind of permanent strike for almost six months" in 2008 (Mikkel Bolt Rasmussen, *Crisis to Insurrection: Notes on the Ongoing Collapse* [Wivenhoe/Brooklyn/Port Watson: Minor Compositions, 2015], 87).

3. Further evidence of this corporate victory is provided by the fact that having suffered an almost ten-year decline, sales of music, films, and video games are now increasing again thanks to digital services such as Apple's iTunes, Amazon's LoveFilm, Spotify, and Netflix. Most evenings Netflix takes up 40 percent of bandwidth in the United States.

4. Felix Stalder, "Enter the Swarm: Anonymous and the Global Protest Movements," *Neural* no. 42 (summer 2012): 7. A time line of events associated with Anonymous is available on Wikipedia at http://en.wikipedia.org/wiki/Timeline_of_events_associated_with_Anonymous, accessed May 22, 2015.

5. To provide an illustrative example, this is Bethany Nowviskie's (mistaken, if oft-cited) argument for adopting CC-BY rather than CC-BY-NC in her blog post, "Why, Oh Why, CC-BY?" Nowviskie writes, "I've concluded that CC-BY is more in line with the practical and ideological goals of the Commons" (Bethany Nowviskie, "Why, Oh Why, CC-BY?" *Bethany Nowviskie,* May 11, 2011, http://nowviskie.org/2011/why-oh-why-cc-by/).

6. See Florian Cramer, "The Creative Common Misunderstanding," *nettime,* October 9, 2006, http://www.mail-archive.com/rohrpost@mikrolisten.de/msg00798.html, republished in Florian Cramer, *Anti-Media: Ephemera on Speculative Arts* (Amsterdam: Institute of Network Cultures, 2013); and Dmytri Kleiner, *The Telecommunist Manifesto* (Amsterdam: Institute of Network Cultures, 2010).

7. Andrew Ross, *Nice Work If You Can Get It: Life and Labor in Precarious Times* (New York: New York University Press, 2009), 168. Further references are contained in the main body of the text.

8. David Golumbia, "Intellectual Property: Hacking Communities and Traditional Knowledge," *empyre,* April 24, 2014.

9. Dmytri Kleiner, "OSW: Open Source Writing in the Network," *empyre,* January 13, 2012, http://www.mail-archive.com/empyre@lists.cofa.unsw.edu.au/msg03634.html.

10. Kleiner, *Telecommunist Manifesto,* 42. Further references are contained in the main body of the text.

It is important to mention in this context another group concerned with preventing uses of their works that are not based in the Commons. Many of those speaking for indigenous communities, their knowledges and traditions, emphasize the need for plurality and exception when it comes to that version of IP law that is underpinned by the logic of the World Intellectual Property Organization and bilateral Free Trade Agreements. This introduction of plurality and exception into debates over the politics of the Commons often involves finding means of withdrawing indigenous or traditional knowledges from the global market and of creating the necessary barriers to prevent their exploitation at the hands of neoliberal corporate capital. See, for one example, the discussion of "The Potential Role of Commons-Based Reciprocity Licenses to Protect Traditional Knowledge" at the FLOK (Free, Live, Open) *Research Plan*, April 23, 2014, http://en.wiki.floksociety.org/w/Research_Plan. Based on an adaptation of Kleiner's proposal for an example of copyfarleft known as a peer production license, these licenses would similarly allow indigenous communities "to freely use the common pool" of knowledges and traditions, but would insist on "a license fee from for-profit companies that want to use the same common pool for the realization of private profit." In this way, they are capable of generating "a stream of income from the private sector companies" in the direction of the indigenous communities and the Commons.

11. For more on the desire of contemporary pirates to escape hierarchies of authority and leadership, see Jonas Staal, ed., in collaboration with Dirk Poot, *New World Academy Reader #3: Leaderless Politics* (Utrecht: BAK, basis voor actuele kunst, 2013). For one of the many available historical accounts of this pirate desire, see the chapter "Hydrarchy: Sailors, Pirates, and the Maritime State," in Peter Linebaugh and Marcus Rediker, *The Many-Headed Hydra: The Hidden History of the Revolutionary Atlantic* (London: Verso, 2000).

12. Roberto Esposito, *Communitas: The Origin and Destiny of Community* (Stanford, CA: Stanford University Press, 2010), 3.

13. Jean-Luc Nancy, *A Finite Thinking* (Stanford, CA: Stanford University Press, 2003), 285.

14. Esposito, *Communitas*, 3. Further references are contained in the main body of the text.

15. For more, see Pauline van Mourik Broekman, Gary Hall, Ted Byfield, Shaun Hides, and Simon Worthington, *Open Education: A Study in Disruption* (New York: Rowman & Littlefield International, 2014). The concern of *Pirate Philosophy* is primarily with the creation, publication, and circulation of knowledge and research. For an engagement with university teaching, however, see this coauthored book.

16. Mark Fisher puts it like this: "I think culture really lags behind politics. I think particularly music it's really glaring—there just doesn't seem to be any music which has substantially grasped the new mood after 2010 really. It seems to me to be a major disjunction between the political situation and cultural forms. The cultural forms that dominate still seem to be so pre-2008 actually. And I think we are really experiencing this lag between culture and politics. Politics is ahead and culture hasn't caught up with it" (Mark Fisher, "Towards a New Hegemony,"

New Left Project, May 17, 2012, http://www.newleftproject.org/index.php/site/article_comments/towards_a_new_hegemony).

17. Rhian E. Jones, *Clampdown: Pop Cultural Wars on Class and Gender* (Winchester: Zero Books, 2013), 89.

18. Samuel Weber, "The Future of the Humanities: Experimenting," *Culture Machine* 2 (2000), http://www.culturemachine.net/index.php/cm/article/viewArticle/311/296; Gary Hall and Clare Birchall, "New Cultural Studies: Adventures in Theory (Some Comments, Clarifications, Explanations, Observations, Recommendations, Remarks, Statements and Suggestions)," in *New Cultural Studies: Adventures in Theory*, ed. Gary Hall and Clare Birchall (Edinburgh: Edinburgh University Press, 2006), 16.

19. David Theo Goldberg, "The Afterlife of the Humanities" (Irvine: University of California Humanities Research Institute, 2014), http://humafterlife.uchri.org/.

20. Does his lack of detailed attention to the specific frames that order and shape critical theory explain why, for all his emphasis on "eschewing imposed and established frameworks," Goldberg's understanding of the power of the humanities today itself often operates within quite conventional humanities and humanist terms and limits. This explains the importance he attaches to "translating the *human* to ourselves: what it is to be, what it means and has meant to be, and what it ought to be *human*" (my emphasis). That the humanist subject still occupies a central place in the world for Goldberg is apparent not least from the approval he displays for Zadie Smith's review of the film *The Social Network*, and with it Facebook, for the *New York Review of Books*. Smith's profoundly liberal humanist analysis offers what Goldberg describes as a "very thoughtful take" on "how digital media generally and social networking applications in particular have changed how we live our lives, communicate and connect, work and play"—of the kind we need more of in the academic humanities, it seems. Zadie Smith, "Generation Why," *New York Review of Books*, November 25, 2010, http://www.nybooks.com/articles/archives/2010/nov/25/generation-why/0.

21. Bernard Stiegler, *Technics and Time, 1: The Fault of Epimetheus* (Stanford, CA: Stanford University Press, 1998), ix.

What is more, this is the case despite the fact that, in terms of the amount of revenue it generates, publishing (i.e., books, journals, newspapers, and magazines) is reputedly the largest of the cultural industries. There are 64,000 publishing companies with "total annual revenues of 121 billion euros" in the European Union alone. Hybrid Publishing Consortium, *Book to the Future—A Manifesto for Book Liberation*, unpublished document, December 2014, 2. These figures are taken from European Commission, "The Commission's Approach to the Publishing Industry: Frequently Asked Questions," European Commission: Press Release Database, September 20, 2005, http://europa.eu/rapid/press-release_MEMO-05-327_en.htm?locale=en. For a range of more recent figures relating to the wider publishing industry in the United States, see Jeffrey R. Di Leo, "Neoliberalism in Publishing: A Prolegomenon," in

Capital at the Brink: Overcoming the Destructive Legacies of Neoliberalism, ed. Jeffrey R. Di Leo and Uppinder Mehann (Ann Arbor, MI: Open Humanities Press, 2014).

22. Nicholas Mirzoeff, "On Hardt and Negri's 'Declaration,'" *Occupy 2012*, May 9, 2012, http://www.nicholasmirzoeff.com/O2012/2012/05/09/on-hardt-and-negris-declaration/.

23. Mark Constantine, cofounder of Lush cosmetics, who has sued Amazon for breach of trademark (although Lush refuses to sell through Amazon, the latter still uses the Lush name to direct consumers to its website before offering them alternative products they might like), describes the business model of the self-styled "everything store" as an instance of "piracy capitalism." He adopts this term to capture the way Amazon "rush into people's countries, they take the money out, and they dump it in some port of convenience. That's not a business in any traditional sense. It's an ugly return to a form of exploitative capitalism that we had a century ago and we decided as a society to move on from" (Mark Constantine, quoted in Carole Cadwalladr, "My Week as an Amazon Insider," *Observer: The New Review*, December 1, 2013, 8–11, 11).

24. http://www.post-crasheconomics.com. While *Pirate Philosophy* is shot through with Marxist and neo-Marxist concerns and is certainly leftist in spirit, its focus is primarily on the human, posthuman, humanities, digital humanities, and posthumanities. For further engagement on my part with neo-Marxist thinkers, including Hardt and Negri, see my "Beyond Marxism and Psychoanalysis," *Culture in Bits* (New York: Continuum, 2002); "Cultural Studies and Deconstruction," in *New Cultural Studies*, ed. Hall and Birchall; and *Digitize This Book! The Politics of New Media, or Why We Need Open Access Now* (Minneapolis: University of Minnesota Press, 2008).

25. This strategy, of developing my argument through careful readings of the most radical and challenging (not to mention celebrated) theorists I can think of, drawn from some of the most important and exciting areas of critical thought today, is one I adopt throughout *Pirate Philosophy*.

26. Graham Harman, "The Importance of Bruno Latour for Philosophy," *Cultural Studies Review* 13 (March 2007): 32. Further references are contained in the main body of the text.

27. Graham Harman, *Prince of Networks: Bruno Latour and Metaphysics* (Melbourne: re.press, 2009). In the interest of full disclosure, I should acknowledge that Harman and Latour edit a series for Open Humanities Press, the open access publisher I cofounded (along with Sigi Jöttkandt and David Ottina) in 2006 and which often uses CC licenses. So, again, it should be stressed that none of this is intended as a moralistic critique of Harman, Latour, re.press, or Creative Commons. For more on Open Humanities Press, see chapter 4.

28. https://creativecommons.org/licenses/by-nc-nd/2.0/.

29. http://creativecommons.org/licenses/by/3.0/.

30. Harman's adherence to a conception of the individual human subject as a legitimate holder of rights and property is not the only example of his humanism. There are others, including the slightly odd emphasis he places on humanizing those associated with object-oriented modes of thought: "Latour is a friendly and approachable figure, a tall man fond of good cigars and good jokes," he tells us (*Prince of Networks*, 11), while Meillassoux is praised for his "friendliness as a colleague." Graham Harman, *Quentin Meillassoux: Philosophy in the Making* (Edinburgh: Edinburgh University Press, 2011). In fact, far from representing a new "third strand" in contemporary thought, object-oriented philosophy is actually a quite traditional philosophy. Nowhere is this more apparent than in its insistence on making the most conventional of philosophical gestures: that which is concerned with announcing a break with tradition and the founding precisely of a new school of thought.

31. Bruno Latour, *We Have Never Been Modern* (Cambridge, MA: Harvard University Press, 1993). Latour's book also comes with a "copyright" and "all rights reserved" statement as part of its inside matter.

Interestingly, Latour's more recent *An Inquiry into Modes of Existence* (Cambridge, MA: Harvard University Press, 2013), which is positioned as offering a positive version of the question raised only negatively in *We Have Never Been Modern*, is accompanied by a "purpose built digital platform allowing for the inquiry summed up in the book to be pursued and modified by interested readers who will act as co-inquirers and co-authors of the final results" (http://www.bruno-latour.fr/node/252, accessed October 8, 2014). All content on this platform, however, is copyrighted on an "all rights reserved" basis too (http://www.modesofexistence.org).

32. Moreover, this state of affairs is not just a problem for theorists of the nonhuman, the inhuman, the posthuman, postanthropocentric, and Anthropocene; it is also an issue for contemporary advocates of both human rights and believers in Darwin's theory of evolution. As Tim Ingold asks: "How can the doctrine of evolutionary continuity be reconciled with the new-found commitment to universal human rights? If, as Article 1 of the Universal Declaration of Human Rights states, all humans are alike in their possession of reason and moral conscience—if, in other words, all humans are the kinds of beings who, according to western juridical precepts, can exercise rights and responsibilities—then they must differ in kind from all other beings which cannot. And somewhere along the line our ancestors must have made a breakthrough from one condition to the other, from nature to humanity" (Tim Ingold, "Beyond Biology and Culture: The Meaning of Evolution in a Relational World," *Social Anthropology* 12 [2004]: 212).

33. See Pirate Cinema Berlin, "Collected Quotations of the Dread Pirate Roberts, Founder of Underground Drug Site Silk Road and Radical Libertarian," *nettime*, October 2, 2013.

34. Daniel Heller-Roazen, *The Enemy of All: Piracy and the Law of Nations* (New York: Zone Books, 2009), 35; "Etymology of Pirate," *English Words of (Unexpected) Greek Origin*.

35. It is impossible to elude such a possibility entirely. Besides, such a dialectical approach may not be an appropriate way of responding to an epistemic environment very different to that known by Marx and Engels, as chapters 3 and 4 show. Rather than directly opposing or negating the unified, sovereign, proprietorial subject, for example, it might be more appropriate to bring this subject into question by exploring some of the tendencies of which it is composed and to mutate and transform it by giving those tendencies new and different inflections.

36. For similar reasons, in addition to experimenting with writing collectively and publishing anonymously in my work, I shift continually between different theories and concepts such as new cultural studies, open media, media gifts, liquid theory, pirate philosophy, radical open access, and disruptive media.

37. Karen Barad, "Posthumanist Performativity: Toward an Understanding of How Matter Comes to Matter," *Signs: Journal of Women in Culture and Society* 28 (2003): 803. Elsewhere, Barad describes the diffractive method of reading as follows: "I call a diffractive methodology, a method of diffractively reading insights through one another, building new insights, and attentively and carefully reading for differences that matter in their fine details, together with the recognition that there [is] intrinsic to this analysis ... an ethics that is not predicated on externality but rather entanglement. Diffractive readings bring inventive provocations; they are good to think with. They are respectful, detailed, ethical engagements" (Karen Barad, "'Matter Feels, Converses, Suffers, Desires, Yearns and Remembers': Interview with Karen Barad," in Rick Dolphijn and Iris van der Tuin, *New Materialism: Interviews and Cartographies* [Ann Arbor, MI: Open Humanities Press, 2012], 50).

38. Elizabeth Grosz, *Time Travels: Feminism, Nature, Power* (Durham, NC: Duke University Press, 2005), 3.

39. Michel Foucault, "What Is Critique?" in *The Politics of Truth,* ed. Sylvère Lotringer and Lysa Hochroth (New York: Semiotext(e), 1997), 45, 47.

40. Judith Butler, "What Is Critique? An Essay on Foucault's Virtue," *Transversal: European Institute for Progressive Cultural Politics* (May 2001), http://eipcp.net/transversal/0806/butler/en.

41. "Critique," Online Etymology Dictionary, http://www.etymonline.com/index.php?term=critique, accessed May 22, 2015.

42. Foucault, "What Is Critique?" 65.

43. Butler, "What Is Critique?"

44. Stanley Fish, "The Digital Humanities and the Transcending of Mortality," *New York Times*, January 9, 2012, http://opinionator.blogs.nytimes.com/2012/01/09/the-digital-humanities-and-the-transcending-of-mortality/.

45. Irit Rogoff, "FREE," in *Are You Working Too Much? Post-Fordism, Precarity, and the Labor of Art*, ed. Julieta Aranda, Brian Kuan-Wood, and Anton Vidokle (Berlin: Sternberg Press, 2011), 199.

Chapter 2

1. Mark Poster, *The Mode of Information: Poststructuralism and Social Context* (Cambridge, UK: Polity, 1990), 147. Such misrecognition is not confined to computer science. As Jacques Derrida points out, a given field cannot understand itself and its founding object because the transcendent position occupied by that object is a "general structure" (109): "The origin of sense makes no sense. ... From this standpoint, technics is not intelligible. This does not mean that it is a source of irrationality, that it is irrational or that it is obscure. It means only that it does not belong, by definition, by virtue of its situation, to the field of what makes it possible. Hence a machine is, in essence, not intelligible. No matter what, even if it makes possible the deployment or transmission or production of meaning, in itself, as machine, it makes no sense" (Jacques Derrida, in Jacques Derrida and Bernard Stiegler, *Echographies of Television* [Cambridge: Polity, 2002], 108–109).

Furthermore, it is a general structure that also applies to humanists and their understanding of the human (see chapter 5). If there is any privilege to be granted the humanities over computer science in this respect, it rests with the fact that while the latter has reflected on its own status as a science and profession (Michael Sean Mahoney, *Histories of Computing* [Cambridge, MA: Harvard University Press, 2011]), the former have traditionally provided the means by which the university thinks of both itself and the identity and relationship of the different professional fields within it. It is a self-questioning role that has been assigned in the UK to English literature and elsewhere to philosophy.

2. See, for example, David M. Berry, "The Computational Turn: Thinking about the Digital Humanities," *Culture Machine* 12 (2011), http://www.culturemachine.net/index.php/cm/article/view/440/470; revised and republished as "Introduction: Understanding the Digital Humanities," in *Understanding Digital Humanities*, ed. David M. Berry (London: Palgrave Macmillan, 2012), 11; danah boyd and Kate Crawford, "Critical Questions for Big Data: Provocations for a Cultural, Technological, and Scholarly Phenomenon," *Information, Communication and Society* 15 (2012): 665.

3. For more on the digital humanities, see chapter 5.

4. Elijah Meeks, "The Digital Humanities as Imagined Community," *Digital Humanities Specialist*, September 14, 2010, http://dhs.stanford.edu/the-digital-humanities-as/the-digital-humanities-as-imagined-community/.

5. National Endowment for the Humanities, Office of Digital Humanities, "Announcing the Digging into Data Challenge," *Digging into Data Challenge*, 2009, http://www.diggingintodata.org/. The Digging into Data Challenge addresses "how 'big data' changes the research landscape for the humanities and social sciences." It does so by "challenging the research community to help create the new research infrastructure for 21st-century scholarship" (National Endowment for the Humanities, *Digging into Data Challenge*, accessed August 5, 2015, http://www.neh.gov/grants/odh/digging-data-challenge).

6. Dan Cohen, "Searching for the Victorians," *Dan Cohen*, October 4, 2010, http://www.dancohen.org/2010/10/04/searching-for-the-victorians/; Lev Manovich, "Trending: The Promises and the Challenges of Big Social Data," in *Debates in the Digital Humanities*, ed. Matthew K. Gold (Minneapolis: University of Minnesota Press, 2012), 467.

7. Although I realize I am going against the tenor of much recent new media theory in saying this, Poster's argument also puts into context the lesson for media studies that some have claimed to be the most lasting legacy of the work of Friedrich Kittler: "His insistence that one needs to have technical understanding of the systems one analyzes and criticizes. In a world where scholars identify with terms like 'digital humanities,' apparently without knowing more than the colloquial meaning of 'digital,' this remains a painfully important message" (Florian Cramer, "Friedrich Kittler," *nettime*, October 29, 2011, http://nettime.org/Lists-Archives/nettime-l-1110/msg00099.html). Yet as Stiegler has pointed out, "instrumental culture cannot be reduced, as it too often is, to the culture of a technician, in the very narrow sense of the word. It is possible to know how to use something without knowing how it works. And it is possible to know how something works and not be able to use it, or to use it only very poorly" (Stiegler, in Derrida and Stiegler, *Echographies of Television*, 57).

8. Jean-François Lyotard (1979), *The Postmodern Condition: A Report on Knowledge* (Manchester: Manchester University Press, 1986). Further references are contained in the main body of the text.

I say "in part" because interestingly, given what I argue below about incommensurability in the humanities, science for Lyotard is a "model of an 'open system'" in that its pragmatics also provide for "dissension," unpredictability and moves that disturb and destabilize the accepted consensus (64, 61)—hence his interest in chaos theory and fractal mathematics, and the emphasis he places at the end of *The Postmodern Condition* on "differential or imaginative or paralogical activity" whose function is to point out "science's 'presuppositions'" and to persuade those involved to "accept different ones" (65). Paralogy, for Lyotard, is thus a form of legitimation "played in the pragmatics of knowledge" (61), admissible because it can "generate ideas" (65), but distinguishable from innovation on the basis that the latter is "under the command of the system, or at least used by it to improve its efficiency" (61). As such, paralogy enables Lyotard to outline a politics that respects both the "desire for the unknown" (67) and "an idea and practice of justice that is not linked to that of consensus" (66).

9. John Houghton, "Open Access—What Are the Economic Benefits? A Comparison of the United Kingdom, Netherlands and Denmark," Centre for Strategic Economic Studies, Victoria University, Melbourne, 2009, http://www.knowledge-exchange.info/Files/Filer/downloads/OA_What_are_the_economic_benefits_-_a_comparison_of_UK-NL-DK__FINAL_logos.pdf.

We also have Houghton's more recently emphasized conclusion, contrary to the UK's Finch report, that while the net benefits of disseminating research by means of gold open access journals would exceed those achievable through green open access repositories if open access were adopted worldwide, given the latter is nowhere near being the case, "the most affordable and cost-effective means of moving towards OA is through green OA" (John Houghton and Alma Swan, "Planting the Green Seeds for a Golden Harvest: Comments and Clarifications on 'Going for Gold,'" *D-Lib Magazine* 19, nos. 1–2 [January—February 2013]).

10. Peter Suber, "Ten Challenges for Open Access Journals," paper presented at the First Conference on Open Access Scholarly Publishing, Lund, Sweden, September 14–16, 2009, http://river-valley.zeeba.tv/ten-challenges-for-open-access-journals//; also available as "Ten Challenges for Open Access Journals," *Sparc Open Access Newsletter*, no. 138, October 2, 2009, http://www.earlham.edu/~peters/fos/newsletter/10-02-09.htm.

11. Alma Swan, "Open Access and Open Data," paper presented at the Second NERC Data Management Workshop, Oxford, UK, February 17–18, 2009, eprints.ecs.soton.ac.uk/17424/.

12. Mark Patterson, "Article-Level Metrics at PloS—Addition of Usage Data," *PLoS Blogs*, September 16, 2009, http://blogs.plos.org/plos/2009/09/article-level-metrics-at-plos-addition-of-usage-data/.

13. Lyotard, *The Postmodern Condition*, 4, 5. To provide another example, Europe PubMed Central, the open access biomedical repository, has linked up with an online system, ResearchFish, for tracking and reporting on research outcomes with a view to connecting "research articles and grant listings with data on their impact": "The idea behind the collaboration is to enable members of the public, researchers and funders to track outcomes and discover which bodies have backed specific articles. The move is also intended to help funders to plan research spending more effectively and highlight the value of the work they support. The information will be publicly available and searchable, with the collaboration so far leading to the pairing of more than 80,000 grants and articles" ("Impact: Research Outcomes Outlined," *Times Higher Education*, June 27, 2013, 12).

14. Roger Burrows, "Living with the H-Index? Metrics, Markets and Affect in the Contemporary Academy," 2011, https://www.academia.edu/807673/Roger

_Burrows_2011_Living_with_the_H-Index_Metrics_Markets_and_Affect_in_the_Contemporary_Academy; published as "Living with the H-Index: Metric Assemblages in the Contemporary Academy," *Sociological Review* 16, no. 2 (2012).

Nor is the altmetrics movement, with its emphasis on impact, an exception to this larger trend. See Jason Priem, Dario Taraborelli, Paul Groth, and Cameron Neylon, "altmetrics: a manifesto," *altmetrics*, October 26, 2010, http://altmetrics.org/manifesto/. More interesting might be a "radical metrics" movement—one that, unlike altmetrics, which is merely a variant of the current conventional metrics, taking into account things like readership numbers, tweets, Facebook shares and "likes," comments, and blogging, would encourage people to work in more interesting ethical and political ways. (My thanks to Rodrigo Costas Comesaña for his input to this idea.) Could we find ways of measuring the amount of intellectual property someone shares or gives away, to offer just one example that is not itself without problems?

15. Jean Kempf, "Social Sciences and Humanities Publishing and the Digital 'Revolution,'" unpublished manuscript, 2010, http://perso.univ-lyon2.fr/~jkempf/Digital_SHS_Publishing.pdf.

16. Lyotard, *The Postmodern Condition*, 6.

17. Sheryl Gay Stolberg, "On First Day, Obama Quickly Sets a New Tone," *New York Times*, January 21, 2009, www.nytimes.com/2009/01/22/us/politics/22obama.html.

18. White House, "Memorandum for the Heads of Executive Departments and Agencies: Transparency and Open Government," January 21, 2009, http://www.whitehouse.gov/the_press_office/TransparencyandOpenGovernment/. For more see the website of the Open Government Initiative: http://www.whitehouse.gov/open.

19. Presidential press secretary Robert Gibbs cited in "White House Condemns WikiLeaks' release," MCNBC.com News, November 28, 2010, http://www.msnbc.msn.com/id/40405589/ns/us_news-security.

20. Dr. Crippen, "Obama's Medical: How Do You Compare?" *Guardian,* G2, February 2, 2010, 2. In her history of transparency, Clare Birchall positions the United States as occupying a leading role in promoting its merits around the world, as "modes of transparency advocated in the U.S. were quickly exported elsewhere" (Clare Birchall, "The Politics of Opacity and Openness," *Theory, Culture and Society* 28 [2011]: 4). And to be sure, the US Democrat equivalent to the http://www.data.gov.uk, www.data.gov, was launched in 2009, six months earlier than the UK site, with the US Freedom of Information Act (FOIA) being implemented in 1966 (as opposed to 2000 in the UK). However, for all it has "been given a modish inflection through its association with and dependency on e-technologies," this emphasis on openness and transparency in government does not originate with the United States. It dates back at least as far as the Enlightenment, a case for open social

communication, as opposed to secret treaties, being made by Immanuel Kant in "Toward Perpetual Peace." Clare Birchall, "Why WikiLeaks Might Not Be as Radical as It Thinks" (paper presented at the Open Media seminar, Coventry School of Art and Design, Coventry University, March 19, 2011), http://coventryuniversity.podbean.com/2011/03/16/why-wikileaks-might-not-be-as-radical-as-it-thinks-clare-birchall/.

21. http://www.un.org/en/documents/udhr/index.shtml.

22. Hillary Clinton, "Remarks on Internet Freedom: The Prepared Text of US Secretary of State Hillary Rodham Clinton's Speech, Delivered at the Newseum in Washington, D.C.," January 21, 2010, http://www.state.gov/secretary/20092013clinton/rm/2010/01/135519.htm.

23. Aaron Swartz, "When Is Transparency Useful?" *Raw Thought*, February 11, 2010, http://www.aaronsw.com/weblog/usefultransparency.

24. Given his administration's pursuit of Snowden, the same can also be said of President Obama's statement at the funeral of Nelson Mandela that "there are too many leaders who claim solidarity with Madiba's struggle for freedom, but who do not tolerate dissent from their own people" (White House, Office of the Press Secretary, "Remarks of President Barack Obama—As Prepared for Delivery," *Remembering Nelson Mandela Johannesburg, South Africa December 10, 2013*, December 10, 2013, http://www.whitehouse.gov/the-press-office/2013/12/10/remarks-president-barack-obama-prepared-delivery).

25. Archon Fung, Mary Graham, and David Weil, *Full Disclosure: The Perils and Promise of Transparency* (Cambridge: Cambridge University Press, 2007), 27–28.

26. Tony Blair, *A Journey: My Political Life* (London: Random House, 2011), 516–517.

27. One of the main influences on the UK's government in this respect is Karl Popper's philosophy of scientific method. For Popper, ideas are true only until they can be proved false. Consequently, he emphasizes the importance of having an open society to enable its ideas to be constantly tested through scientific experimentation as a means of guarding against authoritarianism. See Karl Popper, *The Open Society and Its Enemies*, vol. 1, *The Spell of Plato*; vol. 2, *Hegel and Marx* (London: Routledge, 1945).

In keeping with the neoliberal desire to minimize the role played by the state in society, the UK's government has adopted a variation of Popper's philosophy to justify reforming public services. It has done so on the grounds that it is not just the state that knows how to supply such services—a multiplicity of others do too, including privately owned for-profit businesses. The relevant information and data, including those produced by academic research, therefore need to be made openly available to the public so that ideas of how such services can be provided and funded can be subject to continual testing and experimentation and, indeed, privatization.

28. Frank Pasquale, "The Dark Market for Personal Data," *New York Times*, October 16, 2014, http://www.nytimes.com/2014/10/17/opinion/the-dark-market-for-personal-data.html?_r=0.

29. S. A. Mathieson, "The Public Sector's Data Gold Mine," *Guardian*, December 7, 2011, http://www.guardian.co.uk/public-leaders-network/2011/dec/07/public-sector-data-gold-mine. For more on the care.data project, see Ben Goldacre, "The NHS Plan to Share Our Medical Data Can Save Lives—But Must Be Done Right," *Guardian*, February 21, 2014, http://www.theguardian.com/society/2014/feb/21/nhs-plan-share-medical-data-save-lives.

30. James Manyika, Michael Chui, Brad Brown, Jacques Bughin, Richard Dobbs, Charles Roxburgh, and Angela Hung Byers, "Big Data: The Next Frontier for Innovation, Competition, and Productivity," *McKinsey Global Institute Report* (May 2011).

31. Jean-François Lyotard, *The Inhuman: Reflections on Time* (Cambridge: Polity, 1991), 4.

32. Jeffrey R. Di Leo, "Neoliberalism in Publishing: A Prolegomenon," in *Capital at the Brink: Overcoming the Destructive Legacies of Neoliberalism*, ed. Jeffrey R. Di Leo and Uppinder Mehan (Ann Arbor, MI: Open Humanities Press, 2014), 226.

33. Dominic Rushe, "Facebook's Mobile Move Boost Revenue," *Guardian*, January 29, 2015, 27.

34. In this respect, Amazon, Facebook, Google et al. are making strategic use of the fact that as I make clear in *Digitize Me, Visualize Me, Search Me* (Ann Arbor, MI: Open Humanities Press, 2011, http://www.livingbooksaboutlife.org/books/Digitize_Me,_Visualize_Me,_Search_Me), complete transparency is impossible. This is because there is an aporia at the heart of any claim to transparency. As Clare Birchall shows, "For transparency to be known as transparency, there must be some agency (such as the media [or politicians, or government]) that legitimizes it as transparent, and because there is a legitimizing agent which does not itself have to be transparent, there is a limit to transparency." In fact, the more transparency is claimed, the more the violence of the mediating agency of this transparency is concealed, forgotten, or obscured. Birchall offers the example of "the Daily Telegraph and its exposure of MPs' expenses during the summer of 2009. While appearing to act on the side of transparency, as a commercial enterprise the paper itself has in the past been subject to secret takeover bids and its former owner, Lord Conrad Black, convicted of fraud and obstructing justice" (Clare Birchall, "'There's Been Too Much Secrecy in This City': The False Choice between Secrecy and Transparency in US Politics," *Cultural Politics* 7 [March 2011]: 142). For more on the aporia that lies at the heart of authority, see chapter 5.

35. It is important to note here that Google, Facebook et al. support these governments in turn, enthusiastically backing the Democrats in the United States, for example.

36. http://quantifiedself.com/. For more, see Hall, *Digitize Me*.

37. http://www.suicidemachine.org/.

38. The latter is a stance both Florian Cramer and Alessandro Ludovico insist many students are adopting today. Students are doing so not so much out of characteristic hipster enthusiasm for vintage media (vinyl, cassettes, and so forth), as an attempt to develop forms of social networking that are not controlled by Google, Amazon et al. and their data-mining, monitoring, control, and capitalization of reading. See Florian Cramer, "Afterword," in Alessandro Ludovico, *Post-Digital Print: The Mutation of Publishing since 1894*, Onomatopee 77, 2012, 163. Indeed, Christian Ulrik Andersen and Søren Bro Pold see the urgent challenge for digital literature "in our post-digital print culture" as being "less to get beyond the Gutenberg Galaxy of the printed book" as to "get beyond the Google galaxy of controlled text, the Amazonic textual machinery, the infrastructures of controlled consumption" (Christian Ulrik Andersen and Søren Bro Pold, "Post Digital Books and Disruptive Literary Machines," paper published on the site of Hold the Light: The 2014 ELO Conference, Milwaukee, WI, June 19–21, 2014, 14, http://conference.eliterature.org/sites/default/files/papers/Andersen_Pold_v2_0.pdf).

For more on the strategic withdrawal into inactivity, silence, and passive sabotage, see chapter 5. For one engagement with the paradox inherent in the tendency of many contemporary theorists to work hard at writing and communicating (often quite extensively) about withdrawing from work and the communication networks, see "The No-Work Paradox: An Interview with Vivian Abenshushan by Gabriela Méndez Cota," *Culture Machine* 15 (2014), http://www.culturemachine.net/index.php/cm/article/view/546/554.

39. This is the response Jaron Lanier proposes in *Who Owns the Future?* (New York: Simon & Schuster, 2013).

40. http://i.imgur.com/sIk1QFt.jpg.

41. Gilles Deleuze and Félix Guattari, *A Thousand Plateaus: Capitalism and Schizophrenia* (London: Athlone, 1988), 492.

42. Paul Lewis, "One CCTV Camera for Every 32 People in UK," *Guardian*, March 3, 2011, 5.

43. The right to privacy is one of the three main themes, considered to be nonnegotiable principles, that are shared by all the European political pirate parties. See European Pirate Parties, "The Uppsala Declaration," in Jonas Staal, ed., in collaboration with Dirk Poot, *New World Academy Reader #3: Leaderless Politics* (Utrecht: BAK, basis voor actuele kunst, 2013). For more on sousveillance, see Stuart Armstrong, "How to Get Positive Sousveillance—A Few Ideas," *Practical Ethics*, September 30, 2013, http://blog.practicalethics.ox.ac.uk/2013/09/how-to-get-positive-surveillance-a-few-ideas/.

44. Franco "Bifo" Berardi, *The Soul at Work: From Alienation to Autonomy* (Los Angeles: Semiotext(e), 2009), 121, 120.

45. Jerome Kagan, *The Three Cultures: Natural Sciences, Social Sciences, and the Humanities in the Twenty-First Century* (Cambridge: Cambridge University Press, 2009), 227.

46. As Matthew Kirschenbaum writes, "Whatever else it might be then, the digital humanities today is about a scholarship (and a pedagogy) that is publicly visible in ways to which we are generally unaccustomed. ... Isn't that something you want in your English department?" (Matthew Kirschenbaum, "What Is Digital Humanities and What's It Doing in English Departments?" in Gold, *Debates*, 9).

47. Cohen, "Searching for the Victorians."

48. Patricia Cohen, "Digital Keys for Unlocking the Humanities' Riches," *New York Times*, November 16, 2010, http://www.nytimes.com/2010/11/17/arts/17digital.html?_r=1&hp=&pagewanted=all; Doug Ramsey, "The Happiness of Cities: Do Happy People Take Happy Images?" UC San Diego News Center, April 21, 2014, http://ucsdnews.ucsd.edu/pressrelease/the_happiness_of_cities_do_happy_people_take_happy_images.

49. Federica Frabetti, "Digital Again? The Humanities between the Computational Turn and Originary Technicity" (paper presented at the Open Media seminar, Coventry School of Art and Design, Coventry University, November 9, 2010), coventryuniversity.podbean.com/2010/11/09/open-software-and-digital-humanities-federica-frabetti/. See also Federica Frabetti, "Rethinking the Digital Humanities in the Context of Originary Technicity," *Culture Machine* 12 (2011), http://www.culturemachine.net/index.php/cm/article/view/431, and "Have the Humanities Always Been Digital? For an Understanding of the 'Digital Humanities' in the Context of 'Originary Technicity,'" in Berry, *Understanding the Digital Humanities*.

50. Frabetti, "Digital Again?"

51. Tanner Higgin, "Cultural Politics, Critique, and the Digital Humanities," *Tanner Higgin: Gaming the System*, May 25, 2010, http://www.tannerhiggin.com/cultural-politics-critique-and-the-digital-humanities/; Alan Liu, "Where Is Cultural Criticism in the Digital Humanities?" in Gold, *Debates*, 490–509.

52. Clare Birchall, "The In/Visible/Introduction," in Clare Birchall, *The In/visible* (London: Open Humanities Press, 2011), http://www.livingbooksaboutlife.org/books/The_in/visible.

53. See Stefanie Posavec, "Writing without Words," *Stefanie Posovec*, accessed March 23, 2015, http://www.stefanieposavec.co.uk/writing-without-words; and Posavec's talk at the Post-Digital Scholar Conference, Leuphana, Centre for Digital Cultures, Luneburg, Germany, November 12–14, 2014.

A further example is provided by the Beautiful Science: Picturing Data, Inspiring Insight exhibition at the British Library, London, February 20–May 26, 2014. With regard to theory and philosophy, other examples of the graphing, if not aestheticization, of data are provided by Simon Raper, "Graphing the History of Philosophy," *Drunks&Lamposts*, June 13, 2013, http://drunks-and-lampposts.com/2012/06/13/graphing-the-history-of-philosophy/; and Brendan Griffen, "Graphs of Wikipedia: Influential Thinkers," *Brendan Griffen*, January 3, 2013, http://brendangriffen.com/blog/gow-influential-thinkers/.

54. Peter Wollen, "Introduction," in *Visual Display: Culture beyond Appearances*, ed. Peter Wollen and Lynn Cooke (Seattle, WA: Bay Press, 1995), 9.

55. Bruno Latour, "Why Has Critique Run Out of Steam? From Matters of Fact to Matters of Concern," *Critical Inquiry* 30, no. 2 (2004): 226 . This is one explanation as to why many exponents of the computational turn appear to display little awareness of the research of "critical media scholars (like Matthew Fuller, Wendy Chun, McKenzie Wark and many others) and hacker activists of the past decade; research that has shown again and again how these very formalisms [that is "the 'quantitative' formalisms of databases and programming"] are 'qualitative,' i.e., designed by human groups and shaped by cultural, economical and political interests through and through" (Florian Cramer, "Re: Digital Humanities Manifesto," *nettime*, January 22, 2009, http://www.mail-archive.com/nettime-l@kein.org/msg01331.html). Liu encapsulates the situation as follows:

> Especially by contrast with "new media studies," whose provocateur artists, net critics, tactical media theorists, hactivists, and so on, blend post-1960s media theory, poststructuralist theory, and political critique into "net critique" and other kinds of digital cultural criticism, the digital humanities are noticeably missing in action on the cultural-critical scene. While digital humanists develop tools, data, and metadata critically, therefore (e.g., debating the 'ordered hierarchy of content objects' principle; disputing whether computation is best used for truth finding or, as Lisa Samuels and Jerome McGann put it, "deformance"; and so on) rarely do they extend their critique to the full register of society, economics, politics, or culture. How the digital humanities advances, channels, or resists today's great postindustrial, neoliberal, corporate, and global flows of information-cum-capital is thus a question rarely heard in the digital humanities associations, conferences, journals, and projects with which I am familiar. Not even the clichéd forms of such issues—for example, "the digital divide," "surveillance," "privacy," "copyright," and so on—get much play. (Liu, "Where Is Cultural Criticism," 491)

However, it also suggests that those who have called for the development of a more critically engaged digital humanities, informed by the discussions at *#transformDH: Transformative Digital Humanities: Doing Race, Ethnicity, Gender, Sexuality, Disability and Class in DH*, for example (http://transformdh.org/) and attentive to questions of power, exploitation, social inequality, and environmental destruction—what has been dubbed the "The Dark Side of the Digital Humanities" (http://www.c21uwm.com/digitaldarkside/)—may be missing the point.

56. N. Katherine Hayles, *How We Think: Digital Media and Contemporary Technogenesis* (Chicago: University of Chicago Press, 2012), 33.

57. Brian Croxall, response to Higgin, "Cultural Politics."

58. Scheinfeldt, quoted in Cohen, "Digital Keys."

59. Tom Scheinfeldt, "Where's the Beef? Does Digital Humanities Have to Answer Questions?" *Found History*, May 12, 2010, http://www.foundhistory.org/2010/05/12/wheres-the-beef-does-digital-humanities-have-to-answer-questions/.

60. Scheinfeldt, response to Cohen, "Searching."

61. Bernard Stiegler, "Knowledge, Care, and Trans-Individuation: An Interview with Bernard Stiegler," *Cultural Politics* 6 (2010): 167–168.

62. Meeks, "The Digital Humanities."

63. Tom Scheinfeldt, "Sunset for Ideology, Sunrise for Methodology?" in Gold, *Debates*, 124–126. Of course, methodology can be opposed to ideology and theory in the manner Scheinfeldt has it only if methodology is used as a substitute or synonym for method. Because it is often used in this way in debates around the digital humanities, I have on occasion adopted such usage myself in this chapter. Strictly speaking, however, methodology is the theoretical analysis of the methods that have been or should be applied to a field of study. Consequently, methodology and theory cannot be opposed, as the former is already an example of the latter.

64. Scheinfeldt, response to Cohen, "Searching for the Victorians."

65. Wendy Brown, *Edgework: Critical Essays on Knowledge and Politics* (Princeton, NJ: Princeton University Press, 2005), 4.

Lest this aspect of my analysis appear somewhat unfair, I should stress the ongoing discussion over how the digital humanities are to be defined and understood features a number of critics of the turn toward techniques and methods derived from computer science who have made a case for the continuing importance of the traditional, theoretically informed humanities. See, in their different ways, for example, not just Higgin, "Cultural Politics," and Liu, "Where Is Cultural Criticism," but also Johanna Drucker, "Humanistic Theory and Digital Scholarship," in Gold, *Debates*, 85–95.

For an analysis that draws attention to some of the elements of misrecognition that are in turn to be found in such a traditional, theoretically informed humanism, see what follows, including my conclusion to this chapter, and also Gary Hall, "The Digital Humanities Beyond Computing: A Postscript," *Culture Machine* 12 (2011), http://www.culturemachine.net/index.php/cm/article/view/441/459.

66. Brown, *Edgework*, 4.

67. Hayles positions Manovich's Cultural Analytics as a "frontier of knowledge construction" in the humanities (*How We Think*, 77). Similarly, in a version of "Where Is Cultural Criticism?" presented at the 2011 MLA convention, Liu positions the Cultural Analytics of Lev Manovich and Jeremy Douglass—which he categorizes as new media studies, but which Manovich closely aligns with the digital humanities (Lev Manovich, "Cultural Analytics: Cultural Analytics and Digital Humanities,"

Software Studies Initiative, June 20, 2009, updated December 2012, http://lab.softwarestudies.com/2008/09/cultural-analytics.html)—as treating "digital materials on the scale of corpuses, databases, distributed repositories, and so on—some of the specialties of the digital humanities—[as] ipso facto cultural phenomena" in a manner much of the digital humanities could learn from). Alan Liu, "Where Is Cultural Criticism in the Digital Humanities?" *Alan Liu*, 2011, accessed March 23, 2015, http://liu.english.ucsb.edu/where-is-cultural-criticism-in-the-digital-humanities. The reading that follows draws its account of Cultural Analytics largely from those texts of Manovich that have been referred to most frequently in this context. They date from 2008 on.

68. Stan Schroeder, "The Internet in 2012: 634 Million Websites, 2.4 Billion Users," *Mashable*, January 17, 2013, http://mashable.com/2013/01/17/the-internet-in-2012-634-million-websites-2-4-billion-users-1-3-trillion-google-searches/.

69. Lev Manovich, "Cultural Analytics: Visualizing Cultural Patterns in the Era of 'More Media,'" *Software Studies Initiative,* 2009, http://softwarestudies.com/cultural_analytics/Manovich_DOMUS.doc; published in *DOMUS*, spring 2009.

70. Clive Humby, cited in Neal Pollack, "The Exabyte Revolution," *Wired: UK Edition*, August, 2012, 114.

71. Lev Manovich, "Interview with Lev Manovich for Archive 2020," *Virtueel_Platform,* May 12, 2009, http://virtueelplatform.nl/kennis/analyzing-culture-in-the-21st-century/.

72. Manovich, "Interview with Lev Manovich for Archive 2020."

73. Lev Manovich, "Cultural Analytics: Cultural Analytics and Digital Humanities."

74. Jeremy Douglass, "The Art of Dominant Color in Film," *Software Studies Initiative,* February 27, 2009, http://lab.softwarestudies.com/2009/02/art-of-dominant-color-in-film.html.

75. Lev Manovich, "Cultural Analytics Lectures by Manovich in UK (London and Swansea), March 8–9, 2010," *Software Studies Initiative,* March 8, 2010, http://lab.softwarestudies.com/2010/03/cultural-analytics-lecture-by-manovich.html.

76. boyd and Crawford, "Critical Questions for Big Data," 688. At the Force of Metadata conference held at Goldsmiths College, University of London, November 29, 2008, Manovich acknowledged that his Cultural Analytics approach could be perceived as being structuralist. However, he claimed he preferred to see it as a new empiricism. For him, it is not structuralist because, although he is using data mining to show underlying structures, he then uses this information to focus on specific cultural objects. So Manovich's interest is in the specific rather than in the abstract generalizations he associates with structuralism. Yet it can be argued that this is not enough for his Cultural Analytics to elude the charge of structuralism. Structuralism does not exclude a focus on the specific; rather, it uses the analysis of larger struc-

tures to help interpret the meaning of specific cultural objects, such as Greta Garbo's face or the new model Citroen car in the case of Barthes's *Mythologies* (London: Paladin, 1973).

One can also imagine generating a reading that would make Russian formalism an important key to understanding Manovich's thinking, especially its relation to history, politics, and the social and apparent lack of a recognizable left politics. Alexander Galloway begins to almost offer something of this kind with regard to both Manovich's *The Language of New Media* (Cambridge, MA: MIT Press, 2001) and (following in the footsteps of Mark Tribes's foreword to the latter), Manovich's early posting to the Rhizome mailing list, "On Totalitarian Interactivity." I say *almost*, because of Galloway's anti-intellectual (in the Wendy Brown sense of the term—see chapter 4) insistence on distancing himself from anything that in the Anglo-American world has been labeled as "poststructuralism"—"so thoroughly gauche today" (Alexander R. Galloway, *The Interface Effect* [Cambridge: Polity, 2012], 9). Instead, he prefers to defend Manovich using an oddly retro, historicist/biographical approach that relates Manovich's interest in formalism to his "youth spent in the Soviet Union": "By the early 1930s, Stalin had made socialist realism the only possible style in the Soviet Union. During this period the Russian formalists were criticized for not paying enough attention to social and historical issues, in essence for being apolitical. The power of the Stalinist machine eventually forced many of these formalists to the margins, or worse, into exile or death. Of course Manovich is no exiled enemy of the state, but because of this history he considers it intellectually dangerous to deny questions of form, poetics, and aesthetics" (7).

77. Lev Manovich, in Kevin D. Franklin and Karen Rodriguez'G, "The Next Big Thing in Humanities, Arts and Social Science Computing: Cultural Analytics," *HPC Wire*, July 29, 2008, http://www.hpcwire.com/2008/07/29/the_next_big_thing_in_humanities_arts_and_social_science_computing_cultural_analytics/.

78. It would be interesting to explore the extent to which something similar can be said about the Critical Data Analytics of Richard Rogers. See, for example, his talk, "Dashboards, Social Media Monitoring and Critical Data Analytics" at the Data Power Conference 2015, University of Sheffield, June 22–23, 2015.

79. For one example, see how Tara McPherson describes the T-Races project of David Theo Goldberg and Richard Marciano: "Combining the historical documents of the Home Owners' Loan Corporation (HOLC), a federal agency that helped instantiate practices of redlining in the 1930s, with a Google maps application, the project provides a compelling origin story for ongoing practices of segregation in several Californian and North Carolinian cities" (Tara McPherson, "Some Theses on the Future of Humanities Publishing, Scholarly and Otherwise," *Future Publishing: Visual Culture in the Age of Possibility* Project 5 of the International Association for Visual Culture, 2013).

Another interesting variation on this theme—certainly with regard to the future of critical theory in the twenty-first century—is provided by Version 3 of McKenzie

Wark's *Gamer Theory*—the visualizations. Although there is not space to analyze them here, Wark positions these visualizations as asking, "Are there new ways of perceiving text, or re-imagining text, that can only happen in the networks? Could visualization change not only how we 'read' but how we write?" (McKenzie Wark, *Gamer Theory 3 (Visualizations)*, Institute for the Future of the Book, April 22, 2007, http://www.futureofthebook.org/mckenziewark/visualizations/).

80. Chris Anderson, "The End of Theory: The Data Deluge Makes the Scientific Method Obsolete," *Wired*, June 23, 2008, http://www.wired.com/science/discoveries/magazine/16-07/pb_theory. For a more recent variation on this theme from *Wired*, see Ian Steadman, "Big Data and the Death of the Theorist," *Wired.co.uk*, January 25, 2013, http://www.wired.co.uk/news/archive/2013-01/25/big-data-end-of-theory.

81. Manovich, "Interview with Lev Manovich for Archive 2020."

82. Ibid.

83. boyd and Crawford, "Six Provocations for Big Data" (paper presented at A Decade in Internet Time: Symposium on the Dynamics of the Internet and Society, Oxford Internet Institute, September 21, 2011), http://papers.ssrn.com/sol3/papers.cfm?abstract_id=1926431; published as boyd and Crawford, "Critical Questions for Big Data."

84. Manovich, quoted in Chris Castiglione, "Lev Manovich: Studying Culture with Search Algorithms," *Society of the Query*, November 15, 2009, http://networkcultures.org/query/2009/11/15/lev-manovich-studying-culture-with-search-algorithms/.

85. Manovich, "Interview with Lev Manovich for Archive 2020."

86. Lev Manovich, "UC San Diego—Software Studies Initiative—Lev Manovich," YouTube, January 9, 2009, http://www.youtube.com/watch?v=xtbzVuDqSas&feature=related.

87. Ibid.

88. Edward Shanken, "Lecture Review: Lev Manovich, 'Cultural Analytics,' Paradiso, May 17," Master of Media website, University of Amsterdam, May 26, 2009, http://mastersofmedia.hum.uva.nl/2009/05/26/lev-manovich-cultural-analytics-lecture-at-paradiso/.

89. Even when Cultural Analytics produces results that differ from initial expectations, the cultural analysis offered often seems rather spurious, bordering on the banal:

> Consider a 2D timeline of TIME MAGAZINE covers we created. …
>
> Such mapping is particularly useful for showing variation in the data over time. We can see how color saturation gradually increases as TIME publication reached its peak in 1968. The

range of all values (i.e., variance) per year of publication also gradually increases—but it reaches its maximum value a few years earlier. It is perhaps not surprising to see that the intensity (or "aggressiveness") of mass media as exemplified by the TIME covers gradually rises up to the end of the 1960s as manifested by changes in saturation and contrast. What is unexpected, however, is that since the beginning of the 21st century, this trend is reversed: the covers now have less contrast and less saturation. (Lev Manovich, "What Is Visualization?" *Poetess Archive Journal*, December 20, 2010: 23)

Yet in what sense is the intensity or aggressiveness of mass media exemplified by the increase in intensity of the color saturation on the 4,535 *Time Magazine* covers analyzed? What exactly is the nature of this connection?

A similar charge of spuriousness (not to mention structuralism/formalism) can be leveled even at those projects explicitly positioned by Liu as not just defining "culture in terms of aesthetic or media artifacts," but as using "the initiative's methods to study culture in a recognizably cultural-critical sense" (Liu, "Where Is Cultural Criticism," in Gold, *Debates*, 504, n. 14). See, for instance, the findings of the project on "2008 U.S. Presidential Campaign Ads," to the effect that "McCain's TV ads are more visually aggressive and radical in visual language than Obama's": "Based on surrounding political rhetoric, public opinion and party lines, we might expect that media team for Obama, the younger and popularly labeled more 'dynamic' candidate would design commercial advertisements that showcase this dynamism. Yet, at least for this small sample set, visual analysis reveals otherwise. John McCain's TV ads comparatively more visually radical, at least for the small-scale sample set of this study" (Tara Zepel, "2008 U.S. Presidential Campaign Ads," *Software Studies Initiative*, December 6, 2010, http://lab.softwarestudies.com/2010/12/2008-us-presidential-campaign-ads.html).

90. Manovich, "Cultural Analytics Lectures by Manovich in UK"; Shanken, "Lecture Review."

91. Anne Burdick, Johanna Drucker, Peter Lunenfeld, Todd Presner, and Jeffrey Schnapp, *Digital_Humanities* (Cambridge, MA: MIT Press, 2012), 41.

92. Manovich, in Franklin and Rodriguez'G, "The Next Big Thing."

93. Manovich, "Cultural Analytics: Cultural Analytics and Digital Humanities."

94. Related questions can be raised with regard to Cohen and Gibb's idea of text mining "the 1,681,161 books that were published in English in the UK in the long nineteenth century." Is Victorian literature just that which appears as, and in, books? What theory of Victorian literature, and the book, is being employed here? Pace the Foucault of "What Is an Author?" (Michel Foucault, "What Is an Author?," in *The Foucault Reader*, ed. Paul Rabinow [Harmondsworth: Peregrine, 1986]) and the Derrida of *Spurs* (Jacques Derrida, *Spurs: Nietzsche's Styles* [Chicago: University of Chicago Press, 1979]), what about notes, diary entries, laundry lists, and scraps along the lines of Nietzsche's "I have forgotten my umbrella"? Are they to be considered part of Victorian literature? If not, why not? And of course on what basis can

we allow Google to take the decision as to what counts as Victorian literature, based on what they have and have not digitized, and what they do and do not allow access to?

95. Lev Manovich, "Cultural Analytics: Overview," *Software Studies Initiative*, June 20, 2009, http://lab.softwarestudies.com/2008/09/cultural-analytics.html. As Manovich makes clear, this focus on structural features is at least partly motivated by practical concerns:

> While we are interested in both content and structure of cultural artifacts, at present automatic analysis of structure is much further developed than the analysis of content. For example, we can ask computers to automatically measure gray tone values of each frame in a feature film, to detect shot boundaries, to analyze motion in every shot, to calculate how color palette changes throughout the film, and so on. However, if we want to annotate film's content—writing down what kind of space we see in each shot, what kinds of interactions between characters are taking place, the topics of their conversations, etc., the automatic techniques to do this are more complex (i.e. they are not available in software such as MAT LAB and imageJ) and less reliable. For many types of content analysis, at present the best way to is annotate media manually—which is obviously quite time consuming for large data sets. In the time it will take one person to produce such annotations for the content of one movie, we can use computers to automatically analyze the structure of many thousands of movies. Therefore, we started developing Cultural Analytics by developing techniques for the analysis and visualization of structures of individual cultural artifacts and large sets of such artifacts—with the idea that once we develop these techniques we will gradually move into automatic analysis of content. (Lev Manovich, "How to Follow Global Digital Cultures, or Cultural Analytics for Beginners," in *Deep Search: The Politics of Search Beyond Google*, ed. Felix Stalder and Konrad Becker [Piscataway, NJ: Transaction Publishers, 2009])

Even here, however, the height of Manovich's ambition seems to be automatically analyze the formal features of content.

96. Manovich, "Cultural Analytics: Visualizing Cultural Patterns in the Era of 'More Media'"; Manovich, "How to Follow Global Digital Cultures, or Cultural Analytics for Beginners."

97. In *Meeting the Universe Halfway: Quantum Physics and the Entanglement of Matter and Meaning* (Durham, NC: Duke University Press, 2007), Barad uses the term *intra-action* to refer to the way objects and individuals do not exist outside of and prior to their interaction, but emerge only through their intra-action. According to Barad, "Individuals do not preexist their interactions; rather, individuals emerge through and as part of their entangled intra-relating" (ix). As she puts it elsewhere, for her,

> The idea that beings exist as individuals with inherent attributes, anterior to their representation, is a metaphysical presupposition that underlies the belief in political, linguistic, and epistemological forms of representationalism. Or, to put the point the other way round, representationalism is the belief in the ontological distinction between representations and that which they purport to represent; in particular, that which is represented is held to be independent of all practices of representing.
>
> ... A posthumanist account calls into question the givenness of the differential categories of "human" and "non-human," examining the practices through which these differential bound-

aries are stabilized and destabilized. (Karen Barad, "Posthumanist Performativity," *Signs: Journal of Women in Culture and Society* 28 [2003]: 804, 808)

For more, see the analysis of posthumanism and performative materiality in chapter 4.

98. Manovich, "How to Follow Global Digital Cultures."

99. Friedrich Kittler, *Gramophone, Film, Typewriter* (Stanford: Stanford University Press, 1999), 200.

100. "Our writing tools are also working on our thoughts," as Friedrich Nietzsche puts it in a letter toward the end of February, in F. Nietzsche, *Briefwechsel: Kritische Gesamtausgabe*, ed. Giorgio Colli and Mazzino Montinari (Berlin/New York: de Gruyter, 1975–84), pt. 3, 1: 172; quoted in Kittler, *Gramophone, Film, Typewriter*, 208. In *Discourse Networks 1800/1900* Kittler writes that, "Rather than presenting the subject with something to be deciphered, [writing] makes the subject what it is" (Friedrich Kittler, *Discourse Networks 1800/1900* [Stanford: Stanford University Press, 1992], 196).

101. In other words, it is not just the objects of knowledge that the big data phenomenon and the social theory that goes with it are changing (boyd and Crawford, "Critical Questions for Big Data"); as we see in chapters 3 and 4, it is *us* as human subjects too. Further investigation, for instance, would likely be capable of revealing at least some aspects of data-driven research as falling into the category of biopolitics: "the attempt, starting from the eighteenth century, to rationalize the problems posed to governmental practice by phenomena characteristic of a set of living beings forming a population: health, hygiene, birthrate, life expectancy, race" (Michel Foucault, *The Birth of Biopolitics: Lectures at the Collège De France, 1978–79* [London: Palgrave Macmillan, 2008], 317). After all, thanks to social media, "For the first time, we can follow imaginations, opinions, ideas, and feelings of hundreds of millions of people. We can see the images and the videos they create and comment on, monitor the conversations they are engaged in, read their blog posts and tweets, navigate their maps, listen to their track lists, and follow their trajectories in physical space" (Manovich, "Trending," 461).

102. This mention of truth is not just a reference to Lyotard; it also refers to the requirement of academics to speak "maybe not truth with a capital T, but ... some kind of truth, the best truth they know or can discover [and] to speak that truth to power" (Stuart Hall, "Epilogue: Through the Prism of an Intellectual Life," in *Culture, Politics, Race, and Diaspora: The Thought of Stuart Hall*, ed. Brian Meeks [Miami: Ian Rundle Publishers, 2007], 289–290).

103. Shanken, "Lecture Review."

104. Lev Manovich, "The Practice of Everyday (Media) Life," March 10, 2010, http://www.manovich.net/DOCS/manovich_social_media.doc. See "Metadata, Mon

Amour" for an example of Manovich's concern with efficiency, where he writes: "In short, while computerization made the image acquisition, storage, manipulation, and transmission much more efficient than before, it did not help us much in dealing with its side effects—how to more efficiently describe and access the vast quantities of digital images being generated by digital cameras and scanners, by the endless 'digital archives' and 'digital libraries' projects around the world, by the sensors and the museums" (Lev Manovich, "Metadata, Mon Amour," www.manovich.net, 2002, http://manovich.net/content/04-projects/039-metadata-mon-amour/36_article_2002.pdf).

105. Manovich, in Franklin and Rodriguez'G, "The Next Big Thing."

106. Ibid.; Lev Manovich, "The Practice of Everyday (Media) Life."

107. Eduardo Navas, "Notes on Cultural Analytics Seminar, December 16-17, 2009, Calit2, San Diego," *Remix Theory*, December 29, 2009, http://remixtheory.net/?p=408. This reading is echoed by the definition of cultural analytics provided by Wikipedia: "Cultural Analytics," *Wikipedia*, http://en.wikipedia.org/wiki/Cultural_analytics, accessed January 26, 2011).

108. Manovich, "Trending," 461–462. Further references are contained in the main body of the text.

109. Gary Hall, *Culture in Bits: The Monstrous Future of Theory* (London: Continuum, 2002).

110. Navas, "Notes on Cultural Analytics Seminar."

111. Dan Cohen, "Some Caveats," *Victorian Books: A Distant Reading of Victorian Publications*, November 19, 2010, http://victorianbooks.org/.

112. Wendy Brown, Judith Butler, and Jean Laplanche, in a passage that was collaboratively written, albeit unintentionally so, put it this way: "Theory is not simply different from description; rather, it is incommensurate with description" (Brown, *Edgework*, 80–81).

113. Cohen, "Searching for the Victorians"; Drucker, "Humanistic Theory and Digital Scholarship," 87; Navas, "Notes on Cultural Analytics Seminar."

114. Manovich, "Trending," 469.

115. Drucker, "Humanistic Approaches to the Graphical Expression of Interpretation." *MIT Video*, May 20, 2010, http://video.mit.edu/watch/humanistic-approaches-to-the-graphical-expression-of-interpretation-9596/. For a variation on this theme, see Geert Lovink's argument that "digital humanities, with its one-sided emphasis on data visualization, working with computer-illiterate humanities scholars as innocent victims, has so far made a bad start in this respect. We do not need more tools; what's required are large research programs run by technologically informed theo-

rists that finally put critical theory in the driver's seat. The submissive attitude in the arts and humanities towards the hard sciences and industries needs to come to an end" (Geert Lovink, "What Is the Social in Social Media?," *e-flux* #40, No. 12, 2012, http://www.e-flux.com/journal/what-is-the-social-in-social-media/#_ftn1).

116. Jeffrey Schnapp and Todd Presner, Digital Humanities Manifesto 2.0, 2011, accessed February 19, 2013, http://jeffreyschnapp.com/wp-content/uploads/2011/10/Manifesto_V2.pdf.

117. Burdick et al., *Digital_Humanities*, 135, 82. For more on the misrecognition of the human in the humanities, see n. 1 as well as Hall, "The Digital Humanities beyond Computing," and chapter 5.

118. Homi K. Bhabha, "The Commitment to Theory," in *The Location of Culture* (New York: Routledge, 1994) 28.

119. According to Florian Cramer, the era of the postdigital has already begun. It is "an age where, on the one hand, 'digital' has become a meaningless attribute because almost all media are electronic and based on digital information processing; and where, on the other hand, younger generation media-critical artists rediscover analog information technology" (Florian Cramer, "Post-Digital Writing," *Electronic Book Review*, December 12, 2012, http://electronicbookreview.com/thread/electropoetics/postal; republished in Florian Cramer, *Anti-Media: Ephemera on Speculative Arts* [Amsterdam: Institute of Network Cultures, 2013]). See also his "What Is 'Post-Digital'?," *A Peer-Reviewed Journal About Post-Digital Research* 3: (2014), http://www.aprja.net/?p=1318.

120. Samuel Weber writes:

> To speak of the Humanities, then, is to imply a model of unity based on a certain idea of the *human*, whether as opposed to the *divine* (medieval, scholastic humanism) or to the non-human animal world. ...
>
> The unity of the *uni*versity remains profoundly bound up with the notion of a universally valid essence of the "human," which is the anthropological correlative of the epistemological universalism that resides at the core of the university as an institution. (Samuel Weber, "The Future of the Humanities: Experimenting," *Culture Machine* 2 [2000], http://www.culturemachine.net/index.php/cm/article/view/311/296)

121. Cary Wolfe describes Stiegler in these terms, in the sense of the "prosthetic coevolution of the human animal with the technicity of tools and external archival mechanisms (such as language and culture)" (Cary Wolfe, *What Is Posthumanism?* [Minneapolis: University of Minnesota Press, 2010], xv). For more on Wolfe and posthumanism, see chapter 4.

Chapter 3

1. See, for example, Franco "Bifo" Berardi, "Cognitarian Subjectivation," in *Are You Working Too Much? Post-Fordism, Precarity, and the Labor of Art*, ed. J. Aranda, B.

Kuan-Wood, and A. Vidokle (Berlin: Sternberg Press, 2011); Jodi Dean, *Blog Theory: Feedback and Capture in the Circuits of Drive* (Cambridge: Polity Press, 2010).

2. See, for example, Manuel Castells, *Networks of Outrage and Hope: Social Movements in the Internet Age* (Cambridge: Polity, 2012); Manuel Castells, "In Brazil the Government Is Listening to the Streets," *Adbusters*, July 1, 2013, https://www.adbusters.org/blogs/manuel-castells-brazil.html; Mikkel Bolt Rasmussen, *Crisis to Insurrection: Notes on the Ongoing Collapse* (Wivenhoe/Brooklyn/Port Watson: Minor Compositions, 2015).

3. For one recent account of attempts to develop nonhierarchical forms of political organization and coordination, in this case in relation to the *indignados* movement in Spain, see Simona Levi, "24M: It Was Not a Victory for Podemos, But for the 15M Movement," *Open Democracy*, June 9, 2015, https://opendemocracy.net/can-europe-make-it/simona-levi/24m-it-was-not-victory-for-podemos-but-for-15m-movement.

4. Mark Poster goes even further: of the "major theorists from the 1970s onwards" whose thought has significantly influenced the study of how the media have affected culture, he identifies Michel Foucault, Jacques Lacan, Louis Althusser, Jürgen Habermas, and Judith Butler as being among those who themselves "either paid no attention at all to the vast changes in media culture taking place under their noses or who commented on the media only as a tool that amplified other institutions like capitalism or representative democracy." Poster positions the Deleuze of "Postscript on Control Societies," along with Vilém Flusser, Marshall McLuhan, Jean Baudrillard, Walter Benjamin, Harold Innis, and Hans Magnus Enzensberger as notable exceptions to this general rule (Mark Poster, "An Introduction to Vilém Flusser's *Into the Universe of Technical Images* and *Does Writing Have a Future?*" in V. Flusser, *Does Writing Have a Future?* [Minneapolis: University of Minnesota Press, 2011], xi). Granted, the digital humanities might be said to constitute one example of scholars beginning to take into account some of the implications that changes in the media landscape have for their own ways of creating, performing, and circulating knowledge and research. Yet how many of those who identify with the digital humanities can at this point be thought of as being a major theorist to rank alongside Foucault, Lacan et al. (or indeed would wish to be, given the emphasis in certain parts of the field on moving away from theory and toward methodology and more positivistic, quantitative, and empirical modes of analysis, as we saw in chapter 2)?

5. *Philosophical Transactions* was founded by the Royal Society in London in 1665. However, it did not become peer reviewed until some years later. Even then, there is debate over what exactly is to be understood by peer review. Some are now arguing that over the course of its long history, peer review has often been undertaken by the editors of a journal rather than by independent referees. What we currently understand as peer review did not emerge until the early nineteenth century, with the term *peer review* itself first being used only in 1967. See Aileen Fyfe, "Peer Review:

Not as Old as You Might Think," *Times Higher Education*, June 25, 2015, https://www.timeshighereducation.co.uk/features/peer-review-not-old-you-might-think.

6. Stanley Fish, "The Digital Humanities and the Transcending of Mortality," *New York Times*, January 9, 2012, http://opinionator.blogs.nytimes.com/2012/01/09/the-digital-humanities-and-the-transcending-of-mortality/.

7. Jacques Derrida, *Of Grammatology* (Baltimore, MD: The Johns Hopkins University Press, 1976), 57.

8. Bernard Stiegler, "Derrida and Technology: Fidelity at the Limits of Deconstruction and the Prosthesis of Faith," in *Jacques Derrida and the Humanities*, ed. Tom Cohen (Cambridge: Cambridge University Press, 2001), 239.

9. Bernard Stiegler, "The Discrete Image," in J. Derrida and B. Stiegler, *Echographies of Television* (Cambridge: Polity, 2002), 155.

10. See Mark Hansen, "Media Theory," *Theory, Culture and Society* 23 (2006): 297–306; N. Katherine Hayles, *How We Think: Digital Media and Contemporary Technogenesis* (Chicago: University of Chicago Press, 2012), 30–31. Hansen writes: "What the massive acceleration of the evolution of technics makes overwhelmingly clear is that human evolution is necessarily, and has always been, co-evolution *with* technics. Human evolution is 'technogenesis' in the sense that humans have always evolved in recursive correlation with the evolution of technics" (300).

11. Geoffrey Bennington, "Emergencies," *Oxford Literary Review* 18 (1996): 185.

12. Mark Hansen, '"Realtime Synthesis' and the Différance of the Body: Technocultural Studies in the Wake of Deconstruction," *Culture Machine* 6 (2004), http://www.culturemachine.net/index.php/cm/article/view/9/8.

13. This idea of value conflicts with that "measured through the concept of information and consequently conceived of as calculable," as the "determination of the undetermined." Bernard Stiegler, *Technics and Time, 2: Disorientation* (Stanford, CA: Stanford University Press, 2009), 98.

14. Ibid., 112.

15. Bernard Stiegler, "Suffocated Desire, or How the Cultural Industry Destroys the Individual: Contribution to a Theory of Mass Consumption," *parrehsia* 13 (2011), http://parrhesiajournal.org/parrhesia13/parrhesia13_steigler.pdf.

16. Stiegler, *Technics and Time, 2*, 6. Technics and the human are here joined together in what Simondon refers to as the "transductive" relationship, "a relationship whose elements are constituted such that one cannot exist without the other—where the elements are co-constituents" (2).

17. David Murphy, editor of *Mobile Marketing Magazine*, quoted in David Smith, "Strong Signals for M-Commerce," *E-Commerce*, September 5, 2012, 8.

18. Bernard Stiegler, *For a New Critique of Political Economy* (Cambridge: Polity, 2010), 11, 43.

19. Ian Bogost, "Ian Became a Fan of Marshall McLuhan on Facebook and Suggested You Become a Fan Too," in *Facebook and Philosophy*, ed. D. E. Wittkower (Chicago: Open Court, 2010), 31–32.

20. Derrida, *Echographies*, 58.

21. "About," Unlike Us: Understanding Social Media Monopolies and Their Alternatives, http://networkcultures.org/wpmu/unlikeus/about/, accessed May 27, 2013. 4chan is a precursor to Anonymous. For a brief history of both, see Felix Stalder, "Enter the Swarm: Anonymous and the Global Protest Movements," *Neural* 42 (summer 2012): 6–9. Interestingly, one of the first organized actions on 4chan was against the social networking site Habbo Hotel.

22. Wikipedia contributors, "FreedomBox," Wikipedia, http://en.wikipedia.org/wiki/FreedomBox, accessed September 25, 2012; https://diasporafoundation.org, accessed May 21, 2015; unCloud: Control Your Own Cloud, http://nimk.nl/eng/uncloud-control-your-own-cloud, accessed August 7, 2015; "What … Will You Build with Etherium?" *Etherium*, https://www.ethereum.org/, accessed July 23, 2014. For a more comprehensive list, see prism-break.org.

23. As Hansen makes clear, this is much more of a problem for Stiegler than it is for Derrida. "Because he thinks that *différance* simply is more originary than technics, Derrida can take for granted the possibility for a critical relationship to teletechnologies, albeit one that (in today's real time scenario) is necessarily deferred to the future. For Stiegler, on the other hand, there can be no such transcendental solution, since the possibility for transcendence is itself transductively correlated with technics" (Hansen, "'Realtime Synthesis'").

24. When describing his early thinking, before he encountered the work of Simondon, Stiegler emphasizes that "'the subject' doesn't interest me. … The 'psychic subject' doesn't interest me. The 'national subject' doesn't interest me. Even the 'technical subject' doesn't interest me, if it exists. However, the manner in which processes constitute themselves interests me and I call that … the idiotext: how an idiocy (*idiocie*) in the original sense of the term, that is, a singularity, is constituted; it's a theory of singularity. A theory of singularity that is not a theory of the subject" (Bernard Stiegler, "A Rational Theory of Miracles: On Pharmacology and Transindividuation," *New Formations* 77 [2013]: 166). Later in the same interview, Stiegler puts it in even starker terms: "I have no theory of the subject as point of origin. For me, the subject or subjectivation is something that is produced in an originally heteronomous process" (170).

25. James Leach usefully sums up this cultural tradition and its relation to individual authorship as follows: "The philosophical roots of Euro-American intellectual property law are generally located in Locke's exposition of labour-based individual

ownership rights, in conjunction with a vision of individual creative genius traced to eighteenth-century and early nineteenth-century Romantic authors. The Lockean view imagines the value of art, or any created work, as emanating from the individual, via labour, and entering the artwork through the mechanical process of its creation. Romantic authorship as a model suggests that individual genius transforms ordinary human experiences into extraordinary original art. The artwork, now a detached possession or event, is considered inanimate, perhaps representing, but not containing, the creativity of its producer. ... The relation that is highlighted by such formulations is between artist and created object, and between perceiver and world beyond, not among makers, collaborators, audiences, perceivers, and creations as aspects of each other" (James Leach, "Step Inside: The Politics of [Making] Knowledge Objects," in *The Politics of Knowledge*, ed. Patrick Beart and Fernando Rubio [London: Routledge, 2012], 91).

The construct known as "Stephen Hawking" is perhaps the most obvious contemporary example of how this romantic conception of the subject works to separate the author from those objects and technologies that provide it with a means of expression. For more, see Helene Mialet, *Hawking Incorporated: Stephen Hawking and the Anthropology of the Knowing Subject* (Chicago: University of Chicago Press, 2012).

26. I am here again playing on Fish's argument in "The Digital Humanities." It is interesting to compare Stiegler's attitude in this respect to that of Hélène Cixous with regard to her early texts especially: "I'm speaking here about those first texts that were demoniacal, that I had great trouble bringing myself to sign, of which I said that 'It wasn't me who wrote them.' Even this was a sentence I couldn't use because I couldn't say 'me', it was much too complicated" (Hélène Cixous, in Hélène Cixous and Frédéric-Yves Jeannet, *Encounters: Conversations on Life and Writing* [Cambridge: Polity, 2012], 11). Walter Benjamin is another thinker who expresses uncertainty about "Our own 'I'" (Walter Benjamin, *Early Writings, 1910–1917*, ed. Howard Eiland [Cambridge, MA: Harvard University Press, 2011], 169).

27. Bernard Stiegler, *Acting Out* (Stanford, CA: Stanford University Press, 2009), 39, 52, 54, 72. Oliver Marchart locates another instance of Stiegler's underlying humanism in his account of the misery, illness, and alienation produced by contemporary capitalism: "Obviously, on the premise of an originary default of origin, there can be no positive substance or origin from which one could be 'alienated' in the first place. Consequently, a figure such as Stiegler's idea of primary narcissism, i.e., positive self-love which supposedly is being destroyed by consumer capitalism, comes to serve as just another foundationalist figure of 'non-alienated' origin" (Oliver Marchart, "Antagonism and Technicity: Bernard Stiegler on *Eris*, *Stasis* and *Polemos*," *New Formations* 77 [2013]: 160–161).

28. There is not the space here to analyze the many differences and similarities between the publishing industries—and their relations to the academic systems and

cultural industries—of France and the English-speaking world. For one historical account, see Jeremy Jennings, "The Deaths of the Intellectual: A Comparative Autopsy," in *The Public Intellectual*, ed. Helen Small (Oxford: Blackwell, 2002).

29. Anyone who doubts the power with which such discourses are enforced should listen to "Episode #2: On Tenterhooks, On the Tenure Track," *3620 Podcast*, Annenberg School for Communication, September 17, 2012, http://podcast.asc.upenn.edu/2012/09/up-next-on-tenterhooks-on-the-tenure-track/. That said, and as I show elsewhere, an author's ability to create with computer media is often perceived as giving his or her written work extra authority and intellectual cache. See Gary Hall, "Notes on Creating Critical Computer Media," *Digitize This Book! The Politics of New Media, or Why We Need Open Access Now* (Minneapolis: University of Minnesota Press, 2008).

30. http://www.arsindustrialis.org.

31. Bernard Stiegler, "Bernard Stiegler, Director of IRI (Innovation and Research Institute) at the Georges Pompidou Center, and www2012 Keynote Speaker," *21st International World Wide Web Conference*, Lyon, France, April 16–20, 2012, http://www2012.wwwconference.org/hidden/interview-of-bernard-stiegler.

32. Ibid.

33. Pharmakon.fr, Ecole de philosophie d'Epineuil-le-Fleuriel, includes a philosophy course accessible on the Internet to those who are members of the Ars Industrialis network and a "summer academy" designed to serve as an occasion to "reflect on what a 'technics of self' is in the digital age." See Bernard Stiegler, "A Rational Theory of Miracles: On Pharmacology and Transindividuation," *New Formations* 77 (2013): 173–174.

34. Stiegler, *For a New Critique*, 11.

35. Alexander R. Galloway, "Bernard Stiegler, or Our Thoughts Are with Control," in *French Theory Today: An Introduction to Possible Futures* (New York: Erudio Editions, 2010), 3.

36. Gilles Deleuze, "Postscript on Control Societies," in *Negotiations: 1972–1990* (New York: Columbia University Press, 1995), 177.

37. Michel Foucault, *Discipline and Punish: The Birth of the Prison* (London: Penguin, 1977), 228.

38. Gilles Deleuze, "Control and Becoming," in *Negotiations*, 174. For more on Deleuze's notion of societies of control, see Gary Hall, Clare Birchall, and Pete Woodbridge, "Deleuze's 'Postscript on the Societies of Control,'" *Culture Machine* 11 (2010), http://www.culturemachine.net/index.php/cm/article/view/384/407, from where the account provided here has largely been taken.

39. As Lev Manovich points out, "The largest data sets being used in digital humanities projects are much smaller than big data used by scientists"; in fact none meet the computer industries' definition of "big data," being for the most part work that "can be done on desktop computers using standard software, as opposed to supercomputers." Manovich adds that he expects this gap "will eventually disappear when humanists start working with born-digital user-generated content (such as billions of photos on Flickr), online user communication (comments about photos), user created metadata (tags), and transaction data (when and from where the photos were uploaded)" (Lev Manovich, "Trending: The Promises and the Challenges of Big Social Data," in *Debates in the Digital Humanities*, ed. Matthew K. Gold [Minneapolis: University of Minnesota Press, 2012], 461).

40. See Charles Duhigg, "How Companies Learn Your Secrets," *New York Times*, February 16, 2012, http://www.nytimes.com/2012/02/19/magazine/shopping-habits.html?pagewanted=all; John Naughton, "Big, Bad Tech: How America's Digital Capitalists Are Taking Us All for a Ride," *Observer*, November 23, 2014, 39; Ethan Zukerman, "Me and My Metadata—Thoughts on Online Surveillance," *... My Heart's in Accra*, July 3, 2013, http://www.ethanzuckerman.com/blog/2013/07/03/me-and-my-metadata-thoughts-on-online-surveillance/.

41. Oliver Burkeman, "Reality Check," *Guardian*, March 15, 2011, http://www.tmcnet.com/usubmit/2011/03/15/5376610.htm.

42. Deleuze, "Postscript on Control Societies," 179.

43. Bernard Stiegler, "Relational Ecology and the Digital Pharmakon," *Culture Machine* 13 (2012): 8, http://culturemachine.net/index.php/cm/article/view/464/501.

44. Deleuze, "Control and Becoming," 175.

45. Deleuze, "Postscript on Control Societies," 179.

46. Witness the banner from the student occupation of the executive boardroom on Middlesex University's Trent Park campus, in Anonymous, "Beyond Measure," *Don't Panic, Organize: A Mute Magazine Pamphlet on Recent Struggles in Education*, December 20, 2010, http://www.smashwords.com/books/view/49035. See also the Edu.factory collective, http://www.edu-factory.org.

47. For one reading of Facebook as a factory, see Trebor Scholz, "Facebook as Playground and Factory," in *Facebook and Philosophy*, ed. D. E. Wittkower (Chicago: Open Court, 2010), 241–252.

48. Deleuze, "Postscript on Control Societies," 179–180.

49. Bernard Stiegler, "Suffocated Desire, or How the Cultural Industry Destroys the Individual: Contribution to a Theory of Mass Consumption," *parrehsia* 13 (2011): 52–61, http://parrhesiajournal.org/parrhesia13/parrhesia13_steigler.pdf.

50. Bernard Stiegler, *Technics and Time, 1: The Fault of Epimetheus* (Stanford, CA: Stanford University Press, 1998), 177.

51. Galloway, "Bernard Stiegler," 11–12.

52. And not just our minds, but our bodies too: "Certain organs—the eye, the hand, the brain—must be coordinated for reading and writing to take place, but the entire body must first be trained to sit for long periods of time" (Bernard Stiegler, *Taking Care of Youth and the Generations* [Stanford, CA: Stanford University Press, 2010], 65).

53. Brian Holmes traces the origins of the public university's ruin in the United States to the "invention of the Cohen-Boyer gene-splicing technique in 1973, and its privatisation by Stanford's patent administrator, Niels Reimers" (Brian Holmes, "Disconnecting the Dots of the Research Triangle: Corporatisation, Flexibilisation and Militarisation in the Creative Industries," in *My Creativity: A Critique of Creative Industries*, ed. Geert Lovink and Ned Rossiter [Amsterdam: Institute of Network Cultures, 2007], 185). Roger Brown does something similar for the university in the UK, beginning with "the Thatcher government's decision in November 1979 to end the subsidy for overseas students' fees, and continued with the introduction of research selectivity in 1986, maintenance loans in 1990, the expansion in the number of universities in 1992 and 2004, and top-up and variable fees in 1998 and 2006, respectively" (Roger Brown, "Restore the Equilibrium," *Times Higher Education*, March 28, 2013, 30; see also his book with Helen Carasso, *Everything for Sale? The Marketisation of UK Higher Education* [London: Routledge/Society for Research into Higher Education, 2013]). For what is the most influential variation on this historical narrative, however, see Bill Readings, *The University in Ruins* (Cambridge, MA: Harvard University Press, 1996).

54. Stefan Collini, *What Are Universities For?* (London: Penguin, 2012), 179; Lord Browne, *Securing a Sustainable Future for Higher Education: An Independent Review of Higher Education Funding and Student Finance*, October 2010, 23, https://www.gov.uk/government/uploads/system/uploads/attachment_data/file/31999/10-1208-securing-sustainable-higher-education-browne-report.pdf.

55. Stevphen Shukaitis and Anja Kanngieser, "Cultural Workers Throw Down Yr Tools, the Metropolis Is on Strike," *The Metropolitan Factory: Making a Living as a Creative Worker*, 2012, 55, http://metropolitanfactory.wordpress.com; http://metropolitanfactory.files.wordpress.com/2012/07/cultural-workers-throw-down-yr-tools-the-metropolis-is-on-strike.pdf.

56. Stiegler, *For a New Critique*, 38, 131, n12.

57. It is interesting in respect of this discussion of time that some see the occupied spaces of the Occupy movement as creating what is described as "their own form of time: *timeless time*, a transhistorical form of time, by combining two different types of experience. On the one hand, in the occupied settlements, they live day by day,

not knowing when eviction will come ... and free of the chronological constraints of their previous, disciplined daily lives. On the other hand, in their debates and in their projects they refer to an unlimited horizon of possibilities of new forms of life and community emerging from the practice of the movement. ... It is an emerging, alternative time, made of a hybrid between the now and the long now" (Castells, *Networks of Outrage and Hope*, 223).

58. http://quantifiedself.com.

59. Deleuze, "Control and Becoming," 175.

60. This is one of the reasons David Willetts, UK minister of state for universities and science at the time, was so willing to support a version of gold, author-pays open access, even though there exist many more responsible ways of achieving open access, as I have argued elsewhere. See David Willetts, "Public Access to Publicly-Funded Research," *BIS: Department for Business, Innovation and Skills*, May 2, 2012, http://www.bis.gov.uk/news/speeches/david-willetts-public-access-to-research; Gary Hall, *Digitize This Book!*

The transfer of the responsibility of paying for publication to the individual author (or the author's funding agency or institution) that is achieved by gold author-pays open access is seen by many in the open access community as a typical neoliberal move. It introduces a new set of gatekeepers—be it at funding council, university vice-chancellor, institutional article processing charge (APC) committee, or research dean-level—capable of exercising control (no doubt by means of the kinds of emphasis on accountability, transparency, evaluation, measurement, and centralized data management that has become a feature of neoliberalism's audit culture). At the same time, gold author-pays open access also serves to establish a market for APCs, and thus potentially another means of inflicting debt onto a public institution in the UK to place alongside the imposition of a system of tuition fees in England that because of government miscalculations in the amount students need to pay, and the likelihood that many of the loans it has made to cover the cost of these fees will never be repaid, is predicted to be more expensive than the mechanism for funding universities it replaced.

61. Skeggs adopts this term with reference to reality TV. See Beverley Skeggs, "The Making of Class and Gender through Visualizing Moral Subject Formations," *Sociology* 39 (2005): 965–982.

62. https://www.facebook.com/legal/terms, accessed April 24, 2015. Whereas at one point he was declaring privacy was over, Mark Zuckerberg has recently introduced anonymous log-ins for some of Facebook's new mobile apps. As he puts it in an interview with *Businessweek*, "I don't know if the balance has swung too far, but I definitely think we're at the point where we don't need to keep on only doing real identity things. ... If you're always under the pressure of real identity, I think that is somewhat of a burden" (Brad Stone and Sarah Frier, "Facebook Turns 10: The Mark Zuckerberg Interview," *Businessweek*, January 30, 2014, http://www.businessweek

.com/articles/2014-01-30/facebook-turns-10-the-mark-zuckerberg-interview#p3). However, as Alice E. Marwick makes clear, "having a good reputation, and being trustworthy and authentic, play such an important role" in Web 2.0 culture that aficionados almost invariably use even those technologies that do permit a certain degree of anonymity to create "a single constructed identity leveraged across multiple media types" (Alice E. Marwick, *Status Update: Celebrity, Publicity and Branding in the Social Media Age* [New Haven, CT: Yale University Press, 2013], 192).

This is one of the reasons some have positioned Anonymous and 4chan as being so radical: they allow for the kind of anonymity that is hard to maintain on corporate platforms and their unfiltered, chaotic content makes it difficult to use algorithms to generate predictions based on the mining of data and recognition of patterns.

63. Anne Baron sees the emphasis on attribution and on making work available for free in much contemporary culture (witness the open access movement's adoption of the CC-BY license—see chapter 1) as leading authors to use their own names very much as brands. Thinking of their names as trademarks in this way compensates authors for any loss they may experience as a result of giving away their work, she argues, as these trademarks are still a means of claiming property rights. It is just that it is now reputation, rather than authorial product, that is the main form of production. See Anne Baron, "Property, Publicity and Piracy" (paper presented at the Media Piracy: The Politics and Practices of Borrowing symposium, Queen Mary University of London, June 17, 2014).

64. Maurizio Lazzarato, "The Misfortunes of the 'Artistic Critique' and of Cultural Employment," in *Critique of Creativity: Precarity, Subjectivity and Resistance in the "Creative Industries,"* ed. Gerald Raunig (London: MayFly Books, 2011), 47, http://mayflybooks.org/?page_id=74. See also Berardi, "Cognitarian Subjectivation."

65. Jason Priem, Dario Taraborelli, Paul Groth, and Cameron Neylon, "altmetrics: a manifesto," *altmetrics*, October 26, 2010, http://altmetrics.org/manifesto.

Mez Breeze provides an example of how such entrepreneurship works in an art context, using the example of James Bridle and his attempt to promote the idea of the New Aesthetic (NA) via a panel at the 2012 SXSW, and on Tumblr:

> The more I think about NA, the more I'm inclined to ponder whether Bridle is using it as an adjunct promotional strategy that mimics start-up/entrepreneurial frameworks: grab a manifest-yet-still-edge-worthy-to-some spinable idea, run it through a concept grinder and link it with a delivery system (in this case, the dangling carrot-bait of merging digital concepts with physical that theorists/academics/creatives/intellectuals just can't resist, with high profile figures being drawn to pontification + publicizing). This "debate bait" then actualizes as an emergent discourse with assured (built-in) funding/exposure strategies through clever generation of its own marketing/PR machine—complete with monetisation through conference creation + academic publications/hype/circuit creation—rather than it acting to ideologically frame a legitimately culturally relevant paradigm that highlights "new" corresponding forms of cultural interpretations regarding the fusion of the digital and physical? (Mez Breeze, "The New Aesthetic: Seeing Like Machines," *empyre*, September 13, 2012)

66. For two different examples from the UK, see the Campaign for the Public University, http://publicuniversity.org.uk/, accessed May 24, 2015, and the values and aims of the Council for the Defense of British Universities, http://cdbu.org.uk/?page_id=10, accessed May 24, 2015.

67. Stiegler, *Technics and Time, 2,* 7, 60, 7. Stiegler provides the following explanation of "epochal redoubling": "A text belongs to knowledge to the degree ... that there can be no end-point to its (re-)reading. The after-effect of this textual 'belonging' to knowledge consists of its being always already between constituted programmatic, *epokhal* stablilities, and as what always returns to haunt them (to topple them in order to reconstitute them). If *tekhnē* suspends the programs in force, then knowledge also returns to suspend all stable effects, *tekhnē*'s 'repercussions,' by redoubling them. This is *epokhal redoubling*" (60).

68. Ibid., 59. Indeed, as Stiegler has made clear with regard to his own origins: "The starting point for my work is the question of memory in Plato; more precisely, in what Plato calls 'anamnesis.'" It also lies with "interrogating the relation between anamnesis and hypomnesis; that is, between artificial memory and writing. In fact in the beginning I wasn't studying philosophy, but linguistics and poetics" (Bernard Stiegler, "A Rational Theory of Miracles: On Pharmacology and Transindividuation," *New Formations* 77 [2013]: 164).

69. See Stiegler, *For a New Critique,* 44. Here, as is frequently the case in Stiegler's work—and in marked contrast to the care with which he reads Heidegger, Husserl, Simondon, and others—there is little or no specificity to the analysis of social networking provided, let alone singularity. Instead, Stiegler tends to deal with social networking in the abstract, as a somewhat vague general category, class, genre, or type of media technology, rather than devoting attention to the critical scrutiny of a particular communication platform or, better, environment, such as Facebook, Instagram, or LinkedIn.

Stiegler's recently translated essay on social networks and Facebook is no exception, consisting more of a rigorous philosophical discussion of the concept of friendship and "'friends' ... understood in the sense of 'contacts'" (20) than of Facebook as a specific social networking environment. For example, Stiegler positions digital and social networks as being much the same thing here: "digital, also known as social, networks" (20), he writes at one point. Yet is there not a difference between a (digital) social network such as Facebook and (digital) social media such as Twitter, even though both may be (part of larger) digital networks? See Bernard Stiegler, "The Most Precious Good in the Era of Social Technologies," in *Unlike Us: Social Media and Their Alternatives,* ed. Geert Lovink and Miriam Rasch (Amsterdam: Institute of Networks Cultures, 2012), 20.

70. Stiegler, *Technics and Time, 2,* 9, 141. Actually, Stiegler can be read as maintaining that the politics of memory made possible by writing also depends on the criteria of judgment being prejudged and predecided (in this case by philology): "In the

first chapter of *Disorientation*, 'The Orthographic Age', I explain that the literal prosthetics of orthographic writing constitutes a unique ground of belief—which opens the space of and for politics by providing access to a past that has thus become properly historical. When I read Plato or Heidegger, I do not question the reliability of the already-there. I do not ask: am I certain of having dealt appropriately with the thought of Plato or Heidegger, who are, after all, dead and buried? I believe, and I believe from the outset that I have dealt appropriately with their thoughts, despite the real possibility of typographical errors or interpolations. It is philology's business to establish the authenticity of source materials; once they have been established, I no longer doubt having access, as if I were there, to Plato's or Heidegger's orthographic thoughts, constituted in the very possibility of a certain after-the-fact re-constitution" (Bernard Stiegler, *Technics and Time, 2*, 8). Yet it is precisely these questions regarding originality, authenticity, fixity, and so forth that for me, digital media can help to keep open. See chapter 4 for more.

71. Stiegler, "Bernard Stiegler, Director of IRI"; "Relational Ecology and the Digital Pharmakon," *Culture Machine* 13 (2012): 8, http://culturemachine.net/index.php/cm/article/view/464/501.

72. Stiegler, "Relational Ecology," 5; N. Katherine Hayles, "Hyper and Deep Attention: The Generational Divide in Cognitive Modes," *Profession* 13 (2007); Nicholas Carr, *The Shallows: What the Internet Is Doing to Our Brains* (New York: Norton, 2010).

73. Bernard Stiegler and Irit Rogoff, "Transindividuation," *e-flux journal* 14 (March 2010), http://www.e-flux.com/journal/transindividuation/.

74. Stiegler, *Taking Care*, 130. Elsewhere, Stiegler shows how the "power of writing" can also lead to "a process of *subjectivation*, which is actually a *sujéction*, or submission, leading in turn to de-subjectification and *disindividuation*." Stiegler, "The Most Precious Good," 25.

75. Stiegler, "Discrete Image," 163. Stiegler's privileging of writing and the associated forms and techniques of presentation, debate, and critical attention as a means of analyzing digital media technologies is also apparent in his notion that the digital is "the contemporary form of writing" (Stiegler, "Relational Ecology," 15); and in his very concept of grammatization and "the grammatization process, which allows the discretization of behaviours, gestures, talks, flows and moves of any kind and which consists in a spatialization of time" (Stiegler, "Bernard Stiegler, Director of IRI"). Stiegler's understanding of this process too—for all its association with behaviors, gestures, and moves of any kind—is derived from writing and literacy. Thus, in *Echographies*, Stiegler says, "Just as, in print culture, the school was created to develop this kind of knowledge, we can imagine that a kind of knowledge of the image might be constituted." Yet for him, "it is pretty obvious we can't conceive of grammarians, or therefore of teachers or of students, unless the technics of writing, which gives language this lettered relation, and the instrumental kinds of knowledge it makes possible, are to a large extent appropriated" (58–60). This raises the

intriguing question, might this be interpreted as an instance of the attitude Flusser identifies when he insists:

> Writing, in the sense of placing letters and other marks one after another, appears to have little future. Information is now more effectively transmitted by codes other than those of written signs. ...
>
> Many people deny this out of laziness. They have already learned to write, and they are too old to learn the new codes. (Vilém Flusser, *Does Writing Have a Future?* [Minneapolis: University of Minnesota Press, 2011], 3)

76. Stiegler, "Transindividuation."

77. In suggesting Stiegler has taken insufficient care with regard to his own subjectivity, it would appear at first glance that I am doing so in the Delphic sense of Stiegler not knowing himself, rather than not having taken care of himself. However, as Foucault shows, these two principles of care are not so easy to keep separate. This is certainly the case in the texts of the ancient Greeks and Romans, where it is the need to care for oneself that brings the Delphic maxim of "Know yourself" into operation. Michel Foucault, "Technologies of the Self," in *Technologies of the Self: A Seminar with Michel Foucault*, ed. L. H. Martin, H. Gutman and P. H. Hutton (London: Tavistock, 1988), 20.

78. Stiegler, "Bernard Stiegler, Director of IRI."

79. George Monbiot, "Academic Publishers Make Murdoch Look Like a Socialist," *Guardian*, August 29, 2011, http://www.theguardian.com/commentisfree/2011/aug/29/academic-publishers-murdoch-socialist.

80. There is not the space here to go into the political economy of academic publishing in any great detail. Wilhelm Peekhaus helpfully provides some figures, however:

> Journal prices in the United States increased by 10.8 per cent in 1995, 9.9 per cent in 1996, 10.3 per cent in 1997, and 10.4 per cent in 1998. According to other survey data, the average serial unit cost more than tripled between 1986 and 2003, increasing from US$89.77 to US$283.08. This increase far outpaced the 68 per cent rate of inflation during this same period. ... Even in the most recent years following the global economic meltdown of 2008, serials prices rose at rates between four and five per cent, well above the negative rate of inflation in 2009 and the 1.6 per cent level of inflation in 2010. According to EBSCO Publishing, between 2007 and 2011 journal prices increased by almost 30 per cent for U.S.-based titles and almost 34 per cent for non-U.S. titles. (Wilhelm Peekhaus, "The Enclosure and Alienation of Academic Publishing: Lessons for the Professoriate," *tripleC* 10 [2012]: 582, http://www.triple-c.at/index.php/tripleC/article/view/395/380)

81. In his *Guardian* article Monbiot writes that the "average cost of an annual subscription to a chemistry journal is $3,792. Some journals cost $10,000 a year or more to stock. The most expensive I've seen, Elsevier's *Biochimica et Biophysica Acta*, is $20,930" (Monbiot, "Academic Publishers Make Murdoch"). Meanwhile, Heather Morrison argues that Elsevier's 2009 profits alone could support a global, fully open access scholarly publishing system (Heather Morrison, "Elsevier 2009 $2

Billion Profits Could Fund Worldwide OA at $1,383 per Article," *Imaginary Journal of Poetic Economics*, April 27, 2010, http://poeticeconomics.blogspot.co.uk/2010/04/elsevier-2009-2-billion-profits-could.html). The *Economist* reports Elsevier as making profits of £780m from revenues of £2.1 billion in 2012 ("No Peeking … ," *Economist*, January 11, 2014, http://www.economist.com/news/science-and-technology/21593408-publishing-giant-goes-after-authors-its-journals-papers-no-peeking). Recent research reveals that in the UK, just nineteen universities belonging to the Russell group spent "over £14.4 million (excluding VAT) on subscriptions to journals published by Elsevier alone" (Michelle Brook, "The Cost of Academic Publishing," *Open Access Working Group*, April 24, 2014, http://access.okfn.org/2014/04/24/the-cost-of-academic-publishing/). See also Tim Gower, "Elsevier—the Facts," *Gower's Weblog*, April 24, 2014, http://gowers.wordpress.com/2014/04/24/elsevier-journals-some-facts/.

82. http://thecostofknowledge.com.

83. David Harvie, Geoff Lightfoot, Simon Lilley, and Kenneth Weir, "What Are We to Do with Feral Publishers?" *Organization* 19 (November 2012).

84. According to an analysis published by the *Guardian*, "four US companies—Amazon, Facebook, Google and Starbucks—paid just £30m tax on sales of £3.1bn" in the four years stretching from 2008 to 2012. To put this in context, "collecting the taxes that these companies have wriggled out of would go a long way to shrinking the deficit for which working- and middle-class Britain's living standards are being sacrificed," not to mention funding for universities. Seumus Milne, "A Roll Call of Corporate Rogues Who Are Milking the Country," *Guardian*, October 30, 2012, http://www.guardian.co.uk/commentisfree/2012/oct/30/roll-call-corporate-rogues-tax.

85. Harvie et al., "What Are We to Do." According to Lilley, Harvie et al. came across "companies enjoying profit margins as high as 53 per cent on academic publishing. That compares with 6.9 per cent for electricity utilities, 5.2 per cent for food suppliers and 2.5 per cent for newspapers" (Simon Lilley, "How Publishers Feather Their Nests on Open Access to Public Money," *Times Higher Education*, November 1, 2012, 30–31). Indeed, they identify "only two other industries where these sorts of return are on offer: that in illegal drugs and the delivery of university-level business education" (Harvie et al., "What Are We to Do").

Nor do independent publishers escape their attention. Harvie et al. also call on editors, writers, and readers to abandon *Organization*, the journal in which they published their paper, and start up an identical yet more affordable alternative, if its publisher, Sage, which has an operating profit margin of a little below 19 percent and "gross profit across both books and journals of over 60 per cent," does not lower its prices to those of a comparable society title, "such as the £123 charged for the AMJ or the £182 for ASQ." Here again, the adoption of a similar stance by editors, writers, and readers of critical theory and philosophy would have consequences for some of the most highly respected titles in these fields, including *Theory, Culture and Society* (to provide just one example), for which an institutional print-only subscription (at the time of this writing) is £906.00. Interestingly, the entire editorial board of the *Journal of Library Administration* resigned in 2013 in protest at the author agreement of Taylor & Francis, which included a $2,995 fee to be paid by the author for each open access article. For more, see Brian Matthews, "So I'm Editing This Journal and ... ," *Ubiquitous Librarian*, March 23, 2013, http://chronicle.com/blognetwork/theubiquitouslibrarian/2013/03/23/so-im-editing-this-journal-issue-and/.

86. A full list of Taylor & Francis's journals is available at http://www.tandfonline.com/action/showPublications?display=byAlphabet.

87. Deleuze, "Control and Becoming," 175. As Peekhaus notes in "The Enclosure and Alienation of Academic Publishing," the bundles put together by the megapublishers constitute a considerable percentage of that part of the annual budget a library devotes to acquisitions. Consequently, when a library needs to make cuts in its collection, all too frequently it is "the stand-alone journals from smaller publishers that are cancelled." So not only do they reduce access to the research literature for the academic community that is served by the library, "bundling practices also intensify the tendencies towards concentration and monopoly power of the commercial publishing oligarchs" (583).

88. Jean Kempf, "Social Sciences and Humanities Publishing and the Digital 'Revolution,'" unpublished manuscript, 2010, http://perso.univ-lyon2.fr/~jkempf/Digital_SHS_Publishing.pdf; John Thompson, *Books in the Digital Age: The Transformation of Academic and Higher Education Publishing in Britain and the United States* (Cambridge: Polity Press, 2005).

89. "Greco and Wharton state that the average library monographs purchases have dropped from 1500 in the 1970s to 200–300 currently. Thompson estimates that print runs and sales have declined from 2000–3000 (print runs and sales) in the 1970s to print runs of between 600–1000 and sales of in between 400–500 nowadays." Janneke Adema and Eelco Ferwerda, "Open Access for Monographs: The Quest for a Sustainable Model to Save the Endangered Scholarly Book," *LOGOS* 20, no. 1–4 (2009): 182, n. 10. See also Janneke Adema and Gary Hall, "The Political Nature of the Book: On Artists' Books and Radical Open Access," *New Formations* 78 (summer 2013), https://curve.coventry.ac.uk/open/items/bec7fd48-e138-4bb1-840e-bc664e3e6ca1/1/.

90. Adema and Ferwerda find that "average print runs of 2000 books were quite common" in the 1970s, "whereas at the start of the new century, figures of around 400 copies have become more commonplace" (Adema and Ferwerda, "Open Access for Monographs," 177).

91. Bernard Stiegler, in Bernard Stiegler and Irit Rogoff, "Transindividuation," *e-flux journal* 14, March 2010, http://www.e-flux.com/journal/transindividuation/.

92. In 2008 Striphas was already able to show that "for-profit publishers have a stake in 62% of all peer-reviewed scholarly journals." Given that even a "conservative estimate places the total number of peer-reviewed journals now in existence at around 20,000"—although some have placed it considerably higher—"this means commercial entities as a whole control some 12,400 of them, two-thirds of which they own exclusively," the remaining third being produced "under contract with various learned societies" (Ted Striphas, "'Acknowledged Goods' Worksite," *Differences and Repetitions: The Wiki Site for Rhizomatic Writing,* 2008, http://striphas.wikidot.com/acknowledged-goods-worksite; published as "Acknowledged Goods: Cultural Studies and the Politics of Academic Journal Publishing," *Communication and Cultural/Critical Studies* 7, no. 1 [2010]).

93. Stiegler, "Suffocated Desire."

94. "One particularly potent mechanism of control" for Peekhaus "is the almost universal practice among commercial journal publishers to make publication of scholarly articles contingent upon the author agreeing to transfer the intellectual property rights in a work to the publisher." He notes how this "ability to demand ownership rights in the work of academic labourers has been partly facilitated by a relatively conservative system of tenure and promotion that reinforces the *status quo* of corporate controlled journal venues" (Peekhaus, "The Enclosure and Alienation of Academic Publishing," 581–582). Yet contrary to the impression Peekhaus gives, this practice is far from confined to commercial academic publishers. It applies to the majority of nonprofit, non—open access book publishers, relatively few of which use even a Creative Commons license that would allow authors to retain the rights to their own work. In this respect, given that in the era of smart phones, tablets, and e-readers, reading is increasingly shifting toward electronic media, paper publishing is presented by some as having "largely become a form of Digital Rights Management for delivering PDF files in a file sharing-resistant format" (Florian Cramer, "Post-Digital Writing," *Electronic Book Review,* December 2012, http://electronicbookreview.com/thread/electropoetics/postal).

95. Dmytri Kleiner, *The Telecommumist Manifesto* (Amsterdam: Institute of Network Cultures, 2010), 31.

96. Stiegler, "Relational Ecology," 13.

97. As Kleiner's fellow member of the Telekommunisten Collective, Baruch Gottlieb, writes, "Media Archaeology is thus really a fashion, something inordinately hyped to sell more books, music, clothes, etc. ... The past is being colonized, mined" (Baruch Gottlieb, "Incompatible Research Practices—Week 01—from Functionaries to Programmers [and Then Some Tricks for Handling the Incommensurable]," *empyre,* February 6, 2012). By the same token, "an expression, like Object Oriented

Philosophy can be used to help a thinker/writer/professor acquire the imprimatur of pertinence or authenticity. It obviously helps people trying to capitalize on that particular cultural producer by defining market sector, enhancing the producer's attractiveness to the capitalist" (Baruch Gottlieb, "The Pitfalls of Trendy Theory and the Popular Arts," *empyre*, February 8, 2012).

In this respect it is interesting to read the account of the importance of the "egalitarian nature of blogs" to the development of object-oriented ontology provided by Bryant et al. (Levi Bryant, Nick Srnicek, and Graham Harman, eds., *The Speculative Turn: Continental Materialism and Realism* [Melbourne: re.press, 2011], 6), in the light of Robertson's point that "the blog reveals itself to be the logical conclusion of a process by which each scholar, because of the uniqueness of her scholarship, must establish her own journal in which to publish" (Benjamin J. Robertson, "The Grammatization of Scholarship," *Amodern*, January 16, 2013, http://amodern.net/article/the-grammatization-of-scholarship/). For a brief analysis of how notions of the author, originality, and copyright are enacted in object-oriented philosophy, see chapter 1. For a critical engagement with new materialism, see chapter 4. For one recent attempt to challenge the "current norms of evaluating, commodifying, and institutionalizing intellectual labor" by means of anonymity, see Uncertain Commons, *Speculate This!* (Durham, NC: Duke University Press, 2013). For an earlier attempt, see "This Is a Test," *Culture Machine* 1 (1999).

98. Deleuze, "Postscript on Control Societies," *Negotiations*, 179. For a reading of "'Žižek' the brand" along similar lines, see Jeremy Gilbert, "All the Right Questions, All the Wrong Answers" in *The Truth of Žižek*, ed. Paul Bowman and Richard Stamp (London: Continuum, 2007). Compare this state of affairs to the antileader and anticelebrity ethic of Anonymous, as described by Gabriella Coleman:

> It is key to note that participants do not only wax philosophical about this commitment; they enact it. Participants remind each other with remarkable frequency that one should not behave like a leader, nor seek personal attention in the media, calling the practice "name fagging" or "leaderfagging." If you do "leaderfag," you most certainly will receive a private or public drubbing, and if you have called a lot of attention to yourself, then with a mere keystroke, you might be instantly banished from IRC. ... I was recently witness to just this very act after a participant had been too public about himself to a reporter, an anon who had not even built social capital by putting himself at risk participating in the DDoS attacks. After reading the article where he had been featured, one interlocutor condensed the collective mood in a mere sentence: "Attempting to use all the work that so many have done for your personal promotion is something i will not tolerate." Then he was killed off—exiled from the IRC network. (Gabriella Coleman, "Anonymous—From Lulz to Collective Action," *New Significance*, May 9, 2011, http://www.thenewsignificance.com/2011/05/09/gabriella-coleman-anonymous-from-the-lulz-to-collective-action/)

99. Stiegler, "Suffocated Desire." As far as Stiegler neglecting to pay sufficient attention to certain things to do with his own way of being and doing as a philosopher is concerned, he has recently acknowledged: "I myself am not completely clear regarding what I think of the idea of 'radical free software,' 'creative commons,' 'open source,' the differences between them and their different modalities; I haven't yet

formed a solid view because I think that in order to have a concerted viewpoint one must spend a great deal of time studying carefully the organisational models and questions, which are also the primary questions particularly regarding property and industrial property" (Bernard Stiegler, "A Rational Theory of Miracles: On Pharmacology and Transindividuation," *New Formations* 77 [2013]: 183).

100. Stiegler, "Suffocated Desire."

101. Ibid. Of course, as Stiegler makes clear, when "selection becomes industrial," as it has in the academic publishing industry, where decisions as to what to publish are made increasingly on economic grounds, the effect is to integrate "a vast array of equipment controlled by economically determined calculations that thus from the very beginning attempt to dissolve the undetermined. But because this industrialization ends in the development of différent identities, such a dissolution is not possible. In other words, two indissoluble tendencies confront each other in this transformation. The future consists of their negotiation" (Stiegler, *Technics and Time, 2*, 100).

102. Stiegler, "Bernard Stiegler, director of IRI."

103. Lawrence Liang, "The Man Who Mistook His Wife for a Book," in *Access to Knowledge in the Age of Intellectual Property*, ed. Gaelle Krikorian and Amy Kapczynski (New York: Zone Books, 2010), 283–284. Further references are contained in the main body of the text.

Chapter 4

1. For one expression of this attitude toward open access, see John Bohannon, "Who's Afraid of Peer Review?" *Science*, October 4, 2013, 60–65. For a list of critical responses to Bohanna's article from the open access community, see Mike Taylor, "John Bohannon's Peer-Review Sting against *Science*," *Sauropod Vertebra Picture of the Week*, October 3, 2013, http://svpow.com/2013/10/03/john-bohannons-peer-review-sting-against-science/.

The announcement from the White House Office of Science and Technology Policy is available as Michael Stebbins, "Expanding Public Access to the Results of Federally Funded Research," *Office of Science and Technology Policy*, February 22, 2013, http://www.whitehouse.gov/blog/2013/02/22/expanding-public-access-results-federally-funded-research. Significantly, as of November 15, 2014, only one of the twenty-one agencies required by this directive to develop an open access policy had actually done so. Heather Joseph, "The State of 'Open': Open Access in the US," OpenCon 2014, Washington, DC, November 15, 2014, http://www.slideshare.net/RightToResearch/state-of-oa-us. For the European Commission's position on open access, see European Commission, "Scientific Data: Open Access to Research Results Will Boost Europe's Innovation Capacity," *Press Release Database*, July 17, 2012, http://europa.eu/rapid/press-release_IP-12-790_en.htm?locale=en.

2. http://www.openhumanitiespress.org.

3. Peter Suber, "Ten Challenges for Open Access Journals," lecture at the First Conference on Open Access Scholarly Publishing, Lund, Sweden, September 14–16, 2009, http://river-valley.zeeba.tv/ten-challenges-for-open-access-journals/; also available as "Ten Challenges for Open Access Journals," *Sparc Open Access Newsletter*, no. 138, October 2, 2009, http://www.earlham.edu/~peters/fos/newsletter/10-02-09.htm.

4. Nigel Vincent supplies some figures extracted from what he acknowledges is a far from complete study of submissions to the UK's 2008 Research Assessment Exercise (RAE): "There is a group of disciplines, represented here by English, French and History, in which up to 40% of the outputs in the leading departments are in the form of books with up to a further 25% coming out as book chapters, and where only about a third of the work appears as articles in journals. For disciplines like Philosophy on the other hand some two thirds of the work is in article form, with about a fifth appearing as chapters and a small but still significant percentage as books" (Nigel Vincent, "The Monograph Challenge," in *Debating Open Access*, ed. Nigel Vincent and Chris Wickham [London: British Academy, 2013]).

5. A brief history of publishing at the University of Michigan runs as follows. The University of Michigan, as a smaller institution, found that it could not justify maintaining all of the separate operations that are associated with scholarly publishing (library, repository, academic press, copyright office, and so on). As a result, Michigan University Press ended up partnering with the university's library to form the Scholarly Publishing Office (SPO), which was created in 2001 to service the publication needs not just of University of Michigan faculty, but of any scholar from any institution worldwide and, what is more, to impose no restrictions on the subject matter it published when doing so. The agency MPublishing has since integrated these activities under its purview to cover all aspects of scholarly output, storage, archiving, and dissemination. In its current guise as Michigan Publishing, the specific services offered by the University Library at Michigan include "publishing monographs in print and electronic forms, hosting and publishing journals, with an emphasis on online, open access formats, developing new digital publishing models with the potential to become community portals for wider knowledge sharing" (http://www.publishing.umich.edu/about/our-organization/, accessed October 9, 2014).

OHP's relationship with Michigan Publishing, designed at least in part to show that scholars and libraries can work together to address the urgency in humanities publishing, lasted until 2014, when it was brought to a close by mutual agreement.

6. OHP's first books were, in no particular order: *The Democracy of Objects* (2011) by Levi R. Bryant; *Immersion into Noise* (2011) by Joseph Nechvatal; *Telemorphosis: Theory in the Era of Climate Change*, vol. 1 (2012) edited by Tom Cohen; *Impasses of the Post-Global: Theory in the Era of Climate Change*, vol. 2 (2012) edited by Henry

Sussman; *Terror, Theory and the Humanities* (2012) edited by Jeffrey R. Di Leo and Uppinder Mehan; and *The Cultural Politics of the New American Studies* (2012) by John Carlos Rowe.

7. They include *New Materialism: Interviews and Cartographies* (2012) by Rick Dolphijn and Iris van der Tuin; *Realist Magic: Objects, Ontology, Causality* (2013), by Timothy Morton; *Architecture in the Anthropocene: Encounters among Design, Deep Time, Science and Philosophy* (2013), edited by Etienne Turpin; *Essays on Extinction* by Claire Colebrook: vol. 1, *Death of the Posthuman* (2014), and vol. 2, *Sex after Life* (2014); *Capital at the Brink: Overcoming the Destructive Legacies of Neoliberalism* (2014), edited by Jeffrey R. Di Leo and Uppinder Mehan; and *Minimal Ethics for the Anthropocene* (2014) by Joanna Zylinska.

8. http://openhumanitiespress.org/feedback/; http://www.livingbooksaboutlife.org/.

9. For the initial raw materials that make up *Hacking the Academy*, see http://hackingtheacademy.org. For the print version of the book, see Daniel J. Cohen and Tom Scheinfeldt, eds., *Hacking the Academy: New Approaches to Scholarship and Teaching from Digital Humanities* (Ann Arbor: University of Michigan Press, 2013).

10. Over the years, funding agencies have been responsible for creating and making large amounts of teaching and research content available online in open access repositories and archives (e.g., JISC Content, Europeana). However, for all the time and money that has been spent on doing so, relatively few people are using and reusing these resources. Part of the brief we were given by Jisc was therefore to cluster existing digital resources from repositories and archives in a wide range of subject areas and repackage them in interesting and engaging ways to help increase their use.

11. David Parry, *Ubiquitous Surveillance* (Ann Arbor, MI: Open Humanities Press, 2011).

12. "Usage Statistics and Market Share of WordPress for Websites," *W³Techs*, http://w3techs.com/technologies/details/cm-wordpress/all/all, accessed February 27, 2014.

13. We use our own hosting for the series rather than the existing WikiBooks platform. We do so for four reasons. First, although it is the same base software, WikiBooks runs a limited set of extensions that constrain what is possible, particularly with the type of media that can be embedded. As a result of managing our own server, the project team is free to choose from among a greatly extended set of capabilities and thus take advantage of developments generated by the wider MediaWiki community. For example, we have been able to embed SoundCloud media, as requested by Mark Amerika for the living book he edits in the series on *Creative Evolution*. Second, the WikiBooks community requires certain stylistic conventions—of which NPOV (neutral point of view) is perhaps the best known—that are not necessarily appropriate for scholarly publications in the humanities. Third, as a public project with an established hierarchy of contributors, editors must be vetted by the

existing WikiBooks community through a history of contributions. This would not have been possible within the development time frame of the Living Books About Life project. Finally, the project team prefers that the LiviBL project have its own URL.

14. No doubt this is one reason much of the open access science-related research content in this series consists of links to material elsewhere rather than the embedded republication and reuse of texts in the living books themselves.

15. See Peter Suber on the BBB definition in the *SPARC Open Access Newsletter*, no. 77, September 2, 2004, http://www.earlham.edu/~peters/fos/newsletter/09-02-04.htm, where he also states that two of the three BBB component definitions (the Bethesda and Berlin statements) require removing barriers to derivative works.

16. Janneke Adema and Gary Hall, "The Political Nature of the Book: On Artists' Books and Radical Open Access," *New Formations*, no. 78 (summer 2013): 152. "An examination of the licenses used on two of the largest open access book publishing platforms or directories to date, the OAPEN (Open Access Publishing in Academic Networks) platform and the DOAB (Directory of Open Access Books), reveals that on the OAPEN platform (accessed May 6, 2012) 2 of the 966 books are licensed with a CC-BY license, and 153 with a CC-BY-NC license (which still restricts commercial re-use). On the DOAB (accessed May 6, 2012) 5 of the 778 books are licensed with a CC-BY license, 215 with CC-BY-NC" (Adema and Hall, "The Political Nature of the Book," 153, n. 53).

17. Adema and Hall, "The Political Nature of the Book," 152. What is more, this distinction is often maintained even in cases where authors associated with the digital humanities have begun to open their books to readers a little more—for example, by making their books open to peer commentary at various stages in the writing process. Most famously, McKenzie Wark did this with *GAM3R 7H30RY*, as did Kathleen Fitzpatrick with her book *Planned Obsolescence* (McKenzie Wark, *GAM3R 7H30RY*, Version 1.1, 2006), http://www.futureofthebook.org/gamertheory, published in print with Harvard University Press in 2007; Kathleen Fitzpatrick, *Planned Obsolescence: Publishing, Technology, and the Future of the Academy* (Media Commons Press, 2009), http://mediacommons.futureofthebook.org/mcpress/plannedobsolescence, published in print with New York University Press in 2011). For a critical engagement with Fitzpatrick's *Planned Obsolescence*, see chapter 5.

18. Kenneth Goldsmith, *Uncreative Writing: Managing Language in the Digital Age* (New York: Columbia University Press, 2011), 14–15.

19. In the introduction to their contribution to the series *Symbiosis*, Janneke Adema and Pete Woodbridge go even further, insisting that a "living book is also a symbiotic book. It is a merging and co-habitation of different media-species, a mash-up of text and video, sound and images, pixels and living, material tissue." They thus see the digital medium as making it possible for the book to be infected increasingly with

"foreign (non-textual) elements as it evolves into something different" (Janneke Adema and Pete Woodbridge, *Symbiosis: Ecologies, Assemblages and Evolution* [Ann Arbor, MI: Open Humanities Press, 2011]. http://www.livingbooksaboutlife.org/books/Symbiosis).

20. For more on the capitalist neoliberal nature of Web 2.0, see Dmytri Kleiner and Brian Wyrick, "Info-Enclosure 2.0," *Mute* 3 (January 2007); and for a more recent, if less overtly politically engaged account, see Alice E. Marwick, *Status Update: Celebrity, Publicity, and Branding in the Social Media Age* (New Haven, CT: Yale University Press, 2013).

21. This is not to say the identity and authority of the individual author is necessarily challenged by the use of a wiki. While they do permit emphasis to be placed on working cooperatively and collaboratively—and while individual contributions often have little value or meaning outside the context of the whole—most wikis are nonetheless configured to enable their administrators, if not their editors and users themselves, to attribute responsibility for any particular contribution to an identifiable author or at least to an identifiable IP address. The majority of wikis also provide a page and version history that retains and tracks all individual changes and modifications.

While it is of course the largest and most successful example of a collaborative wiki project, even Wikipedia does not escape the authority of the author. In addition to the factors detailed above, authors also need a relatively high standard of writing or technical skills to have a realistic chance of being able to contribute to Wikipedia. So it is not the case that just anyone can do so. A form of authorial authority also resides in the figure of the Wikipedia moderator, who is able to resolve any disputes over the contents of entries and so on. Moreover, the means of resolving such conflict on Wikipedia is ostensibly based on rationality (and this includes the discussion pages associated with each Wikipedia entry in which people can set out, explain, and justify the reasoning behind any changes they have made). This practice is founded in turn on much the same notion of the rational human individual as that which underpins the romantic, possessive model of the author.

Interestingly, however, Wikipedia has become too large to be written and managed by humans alone. Consequently, its information and updates are being generated increasingly by bots that take information from one place and post them to another, or port data from a version of Wikipedia in one language to a version in a different language, sometimes using a related intermediary such as Wikidata. Thomas Steiner has created an open source application (http://wikipedia-edits.herokuapp.com/) that illustrates in real time just how many Wikipedia entries are being written by humans and how many by bots.

Nor is it the case that wikis cannot be articulated in liberal, neoliberal, or libertarian terms. Far from it. Wikia (http://www.wikia.com/About), for example, is a for-profit enterprise from the Hayekian-influenced founder of Wikipedia, Jimmy Wales, which offers free Web hosting for wikis. Wikia is free to readers and editors. It makes money by selling its communities to advertisers, helped by the brand recognition of Wikipedia, from which it is understandably careful to stress it remains separate as a business, although both share MediaWiki technology.

22. See http://punctumbooks.com and http://geologicnow.com/; and Sönke Bartling and Sascha Friesike, eds., *Opening Science: The Evolving Guide on How the Web Is Changing Research, Collaboration and Scholarly Publishing*, http://book.openingscience.org, accessed December 9, 2014.

23. http://liquidbooks.pbworks.com/w/page/21175721/Technology%20and%20Cultural%20Form%3A%20A%20Liquid%20Reader. As an example of what else can be achieved, a volume in the Liquid Books series, *Biomediaciones/Biomediations*, was created in only three hours by a group of fifteen collaborators at the 2013 Festival of New Media Art and Video, Transitio_MX 05 BIOMEDIATIONS (Biomediaciones), Mexico City, http://liquidbooks.pbworks.com/Volume%206%20Biomediaciones%20%20Biomediations.

24. Vincent, "The Monograph Challenge." Having been commissioned by the Higher Education Funding Council for England to explore issues surrounding open access monographs and other long, scholarly works, Geoffrey Crossick, Distinguished Professor of Humanities at the School of Advanced Study, University of London, also takes care to insist that readers should not be able to interact with monographs on a read/write/rewrite basis as they can with Wikipedia. See Holly Else, "Digital Age of Opportunity for the Open Access Monograph," *Times Higher Education*, January 16, 2014, http://www.timeshighereducation.co.uk/news/digital-age-of-opportunity-for-the-monograph/2010476.article.

25. Neil Badmington, "Cultural Studies and the Posthumanities," in *New Cultural Studies: Adventures in Theory*, ed. Gary Hall and Clare Birchall (Edinburgh: Edinburgh University Press, 2006).

26. Rosi Braidotti, *The Posthuman* (London: Polity, 2013), 4. Further references are contained in the main body of the text.

27. Jean-François Lyotard, *The Postmodern Condition: A Report on Knowledge* (Manchester: Manchester University Press, 1986). See chapter 2 for more.

28. Wiley is included in the 40 percent of publishers that embargo green open access (OA), thus "forcing authors," in Stevan Harnad's words, "to pay for hybrid Gold OA as the only way to provide immediate OA to their articles" (Stevan Harnad, "Beyond Double-Dipping: Free-Choice Fair Gold vs. Forced-Choice Fool's Gold," BOAI e-mail list, October 26, 2013). In response to years of above-inflation increases in the cost of its journals, the University of Montreal not so long ago took the step of canceling its subscription to over one thousand Wiley-Blackwell journals that the publisher had been selling to the institution as a "big deal" bundling strategy, which requires institutional libraries to buy large numbers of publisher-generated packages

of journals (Paul Jump, "Montreal Tells Wiley: 'No Deal,"' *Times Higher Education*, January 30, 2014, 13).

29. Peter Williams, Iain Stevenson, David Nicholas, Anthony Watkinson, and Ian Rowlands, "The Role and Future of the Monograph in Arts and Humanities Research," *Aslib Proceedings*, 61 (2009): 67–82, 67.

30. Vincent, "The Monograph Challenge."

31. Bruno Latour, *The Pasteurization of France* (Cambridge, MA: Harvard University Press, 1988).

32. Cary Wolfe, quoted by Braidotti, *The Posthuman*, 1. This is somewhat ironic, given that Braidotti quotes N. Katherine Hayles to the effect that "what is lethal" about biogenetic capitalism "is not the posthuman as such but the grafting of the posthuman onto a liberal humanist view of the self" (N. Katherine Hayles, *How We Became Posthuman: Virtual Bodies in Cybernetics, Literature and Informatics* [Chicago: University of Chicago Press, 1999], 286, quoted by Braidotti, *The Posthuman*, 101).

33. For more on the contradiction involved in a singular author producing a book with his or her name on it about the problems involved in singular authors producing books with only their names on them, and about how my argument in *Pirate Philosophy* relates to the construct known as *Gary Hall-the-theorist* I am enacting here, see in particular chapter 6 and the discussion of diffraction and the art of critique at the end of chapter 1.

34. Stefan Herbrechter and Ivan Callus, "What Is a Posthumanist Reading?" *Angelaki* 13 (April 2008): 105. Further references are contained in the main body of the text.

35. Francis Fukuyama, *Our Posthuman Future: Consequences of the Biotechnology Revolution* (New York: Saint Martin's Press, 2003). Witness Cary Wolfe's reliance on animal studies as a lens through which to embark on his discussion of the "(Post) Humanities" in *What Is Posthumanism?* Cary Wolfe, '"Animal Studies,' Disciplinarity, and the (Post)Humanities," in his *What Is Posthumanism?* (Minneapolis: University of Minnesota Press, 2010).

36. "Posthumanities," University of Minnesota Press, http://www.upress.umn.edu/book-division/series/posthumanities, accessed March 28, 2014.

37. Wolfe, *What Is Posthumanism?* 123, 106. Further references are contained in the main body of the text.

38. Braidotti, *The Posthuman*, 79. Further references are contained in the main body of the text.

Another example of the reinstatement of humanism and the humanities by those purportedly advocating a posthumanities is provided by David Theo Goldberg's "The Afterlife of the Humanities," which I refer to in chapter 1. David Theo

Goldberg, "The Afterlife of the Humanities" (Irvine: University of California Humanities Research Institute, 2014), http://humafterlife.uchri.org/. Goldberg argues for a posthumanities as a means of posing "alternative modalities for taking up, for doing, for engaging (and for an engaging) humanities." Yet for all his emphasis on "eschewing imposed and established frameworks," Goldberg's understanding of the power of the humanities today operates within quite conventional humanist and humanities frames.

39. "Posthumanities," University of Minnesota Press, http://www.upress.umn.edu/book-division/series/posthumanities, accessed March 28, 2014. The longer version of this text Braidotti refers to and quotes from in *The Posthuman* is no longer available at the online address supplied in her bibliography.

40. Braidotti, *The Posthuman*, 1.

41. This is according to the book's index: Wolfe is also cited on 70.

42. I say *so-called* because poststructuralism is not a label many of those who are most often associated with it—including Michel Foucault, Jean-François Lyotard, and Jacques Derrida—either used or felt comfortable with. It is a designation they tended to associate with the take-up of their writings in North America, something that said much about the power structures of the academy there.

43. An example of this most traditional of gestures (that which is concerned to announce a break with tradition) is provided by Peter Sloterdijk's positioning of Jacques Derrida and Niklas Luhmann as being concerned with "perfecting and retouching the finished image of a tradition that could not be extended any further. ... In the case of Derrida, this involved the conclusion of the linguistic or semiological turn according to which the twentieth century had belonged to the philosophies of language and writing" (Peter Sloterdijk, *Derrida, An Egyptian: On the Problem of the Jewish Pyramid* [Cambridge: Polity Press, 2009], 2, 3). The "Semiotic Turns" section in Bruno Latour's *We Have Never Been Modern* (Cambridge, MA: Harvard University Press, 1993), provides another.

44. For one of many possible versions, in addition to that already provided by Braidotti in this chapter and by myself elsewhere in this book, see Jussi Parikka's comments regarding his take on "Zombie Media." Parikka uses the term *zombie* to refer to obsolete, forgotten, dysfunctional, or discarded physical media objects or things that nevertheless continue to exist as a form of "living dead" media through their leaking of material toxins into the natural environment and so forth. Explaining his attempt with Garnet Hertz to "address the planned obsolescence of media technologies which is part of their material nature" through the active repurposing of "things considered dead—things you find from your attic, the second hand market, or amongst waste" (Garnet Hertz and Jussi Parikka, "Five Principles of Zombie Media," *DeFunct/ReFunct*, South Dublin Arts Centre, Dublin, RUA RED,

2011, http://conceptlab.com/writing/hertz-parikka-2011-defunct-refunct-catalogue.pdf; also available as "Zombie Media in Leonardo" on Parikka's *Machinology*, September 5, 2012, http://jussiparikka.net/2012/09/05/zombie-media-in-leonardo/), Parikka writes: "I am interested in politics that is not critique ... something that proceeds through an imminent relation to the technological sphere in which we function (whether as theorists, artists, other kinds of practitioners, or just plainly ordinarily people, consumers). Not critique as standing-against, as a lot of our technology/capitalism critique has been (naturally, Adorno, and a bunch of other technophobes, even if I realize [*sic*] making a caricature here); not critique as opposing, but retwisting, bending, modulating" (Jussi Parikka, Comments on his "Zombie Media: Media Archaeology as Circuit Bending," Open Media seminar series, Coventry University, January 25, 2011, http://coventryopenmedia.wordpress.com/2011/01/07/tuesday-25th-january/#comments). Garnet Hertz and Jussi Parikka, "Zombie Media: Circuit Bending Media Archaeology into an Art Method," a longer version of "Five Principles of Zombie Media," appeared in *Leonardo* 45, no. 5 (2012), and as the Appendix to Jussi Parikka, *A Geology of Media* (Minneapolis: University of Minnesota Press, 2015).

45. Franco "Bifo" Berardi, *The Soul at Work: From Alienation to Autonomy* (Los Angeles: Semiotext(e), 2009), 154.

46. Levi Bryant, Nick Srnicek, and Graham Harman, eds., *The Speculative Turn: Continental Materialism and Realism* (Melbourne: re.press, 2011), http://re-press.org/books/the-speculative-turn-continental-materialism-and-realism.

This rhetorical approach of "no longer"—a version of what in a less theoretical context Evgeny Morozov dubs "epocholism" (Evgeny Morozov, *To Save Everything, Click Here: Technology, Solutionism, and the Urge to Fix Problems That Don't Exist* [London: Allen Lane, 2013]—is itself criticized from within new materialism, even as the latter repeats such reductive clichés. Manual DeLanda writes: "We should not attempt to build such a [new or neomaterialist] philosophy by 'rejecting dualisms' or following any other meta-recipe. The idea that we know already how all past discourses have been generated, that we have the secret of all past conceptual systems, and that we can therefore engage in meta-theorizing based on that knowledge is deeply mistaken. And this mistake is at the source of all the idealisms that have been generated by postmodernism" (Manuel DeLanda, "'*Any Materialist Philosophy Must Take as Its Point of Departure the Existence of a Material World That Is Independent of Our Minds*': Interview with Manuel DeLanda," in Rick Dolphijn and Iris van der Tuin, *New Materialism: Interviews and Cartographies* [Ann Arbor, MI: Open Humanities Press, 2012], 43–44).

47. Although much of what I have to say here in relation to posthumanism and new materialism is based on Deleuze's notion of vitalism, I am concentrating mainly on the materialist aspect of the posthuman subject. For one analysis of vitalism along similar lines, see Dennis Bruining, "A Somatechnics of Moralism: New Materialism or Material Foundationalism," *Somatechnics* 3, no. 1 (2013), 149–168.

48. Stefan Herbrechter, "The Roar on the Other Side of Silence ... or, What's Left of the Humanities?" *Culture Machine Reviews* (2013): 4, http://www.culturemachine.net/index.php/cm/article/viewFile/495/516.

49. Ibid., 5.

50. Gary Hall and Clare Birchall, "New Cultural Studies: Adventures in Theory (Some Comments, Clarifications, Explanations, Observations, Recommendations, Remarks, Statements and Suggestions)," in *New Cultural Studies*, 13, quoting Wendy Brown, *Politics Out of History* (Princeton NJ: Princeton University Press, 2001), 29, 30.

51. Bruining, "A Somatechnics," 151. The new materialist works Bruining engages with include Stacy Alaimo and Susan Hekman, eds., *Material Feminisms* (Bloomington: Indiana University Press, 2008); Diana Coole and Samantha Frost, eds., *New Materialisms: Ontology, Agency, and Politics* (Durham, NC: Duke University Press, 2010); and Susan Hekman, *The Material of Knowledge: Feminist Disclosures* (Bloomington: Indiana University Press, 2010). Papoulias and Callard identify a similar afoundational foundationalism with regard to the empirical-experimental biological evidence that is used to underpin the materialist approach to the theory of affect, as when "Teresa Brennan asserts that 'experiments *confirm* that the maternal environment and olfactory factors ... shape human affect,' and Brian Massumi reassures us that 'the time-loop of experience has been experimentally *verified*.' Even as affect theory shows how a biology of afoundational foundations can be imagined, the language through which the findings of neuroscience are invoked by cultural theorists is, paradoxically, often the language of evidence and verification, a language offering legitimation through the experimental method. It is through the old foundational language, in other words, that the afoundational biology is appropriated" (Constantina Papoulias and Felicity Callard, "Biology's Gift: Interrogating the Turn to Affect," *Body and Society* 16, no. 29 [2010]: 37).

52. Braidotti, *The Posthuman*, 5.

53. Brown, *Politics Out of History*, 29.

54. Federica Frabetti, *Software Theory* (London: Rowman and Littlefield International, 2014).

55. At least one of these series is explicitly associated with new materialism: Rick Dolphin and Iris van der Tuin's *New Materialism: Interviews and Cartographies* appears as part of New Metaphysics. Dolphin and van der Tuin's book includes interviews with Manuel DeLanda, Karen Barad, and Quentin Meillassoux, as well as Braidotti herself. Braidotti is also a member of the editorial board of OHP's Critical Climate Change series.

56. Jacques Derrida, *Spurs: Nietzsche's Styles* (Chicago: University of Chicago Press, 1979).

57. Jacques Derrida, "The Word Processor," in *Paper Machine* (Stanford: Stanford University Press, 2005).

58. Jacques Derrida, "Signature Event Context," in *Margins of Philosophy* (London: Harvester Wheatsheaf, 1982).

59. Jacques Derrida, "Paper, or Me, You Know … (New Speculations on a Luxury of the Poor)," in *Paper Machine* (Stanford: Stanford University Press, 2005).

60. Jacques Derrida, "Freud and the Scene of Writing," in *Writing and Difference* (London: Routledge and Kegan Paul, 1978). Significantly, Derrida writes here that "the 'subject' of writing does not exist if we mean by that some sovereign solitude of the author. The subject of writing is a *system* of relations between strata: the Mystic Pad, the psyche, the society, the world" (226–227).

61. Jacques Derrida, *Of Grammatology* (Baltimore, MD: Johns Hopkins University Press, 1976); "The Book to Come," in *Paper Machine* (Stanford: Stanford University Press, 2005).

62. Jacques Derrida, "Limited Inc," *Glyph* 2 (1977), 162–256.

63. Jacques Derrida, "'This Strange Institution Called Literature': An Interview with Jacques Derrida," in *Acts of Literature*, ed. Derek Attridge (London: Routledge, 1992); "Mochlos; or, The Conflict of the Faculties," *Logomachia: The Conflict of the Faculties*, ed. Richard Rand (Lincoln: University of Nebraska, 1992); "The Principle of Reason: The University in the Eyes of Its Pupils," *Diacritics* 13 (fall): 1983, 2–20; *Who's Afraid of Philosophy? Right to Philosophy* (Stanford, CA: Stanford University Press, 2002); *Eyes of Philosophy: Right to Philosophy II* (Stanford, CA: Stanford University Press, 2004); "The Future of the Profession or the University without Condition (Thanks to the "Humanities" *What Could Take Place* Tomorrow)," in *Jacques Derrida and the Humanities: A Critical Reader*, ed. Tom Cohen (Cambridge: Cambridge University Press, 2001).

64. Jacques Derrida, *Dissemination* (London: Athlone, 1972); *The Post Card: From Socrates to Freud and Beyond* (Chicago: University of Chicago Press, 1980); *Glas* (Lincoln: University of Nebraska Press, 1974), "Tympan," in *Margins of Philosophy* (London: Harvester Wheatsheaf, 1972); "Circumfession," in *Jacques Derrida*, ed. Geoffrey Bennington and Jacques Derrida (Chicago: University of Chicago Press, 1993).

65. Vicki Kirby, *Quantum Anthropologies: Life at Large* (Durham, NC: Duke University Press, 2011), x. Further references are contained in the main body of the text.

"Materialist" is how Karen Barad describes Kirby's reading of Derrida. Karen Barad, "Posthumanist Performativity: Toward an Understanding of How Matter Comes to Matter," *Signs: Journal of Women in Culture and Society* 28 (2003): 829, n. 38.

66. John Protevi, *Political Physics: Deleuze, Derrida and the Body Politics* (London: Athlone, 2001), 19.

67. Kirby, *Quantum Anthropologies*, ix.

68. Tim Ingold, "Materials against Materiality," in *Being Alive: Essays on Movement, Knowledge and Description* (London: Routledge, 2011), 20. Further references are contained in the main body of the text.

69. Another reason for doing so is in order not to contradict what I say above about learning from Derrida the importance of approaching texts hospitably and responsibly so as to avoid slipping into the anti-intellectual moralism that can be detected in many theories of materialism.

70. Johanna Drucker, "Understanding Media: Craig Dworkin's 'No Medium,'" *Los Angeles Review of Books*, July 9, 2013, http://lareviewofbooks.org/review/understanding-media-craig-dworkins-no-medium.

71. Ibid.

72. Erkki Huhtamo and Jussi Parikka, "Introduction: An Archaeology of Media Archaeology," in *Media Archaeology: Approaches, Applications, and Implications*, ed. Erkki Huhtamo and Jussi Parikka (Berkeley: University of California Press, 2011), 8.

73. It should be noted that this reading of media archaeology is not confined to Drucker. To provide one further example, if we pursue Ingold's earlier reference to wasps and bees, we find similar problems to those he identifies in the study of materiality in a book from Wolfe's Posthumanities series, Jussi Parikka's *Insect Media: An Archaeology of Animals and Technology* (Minneapolis: University of Minnesota Press, 2010). This book is often cited as an exemplary instance of contemporary media archaeology. See, for example, Huhtamo and Parikka, "Introduction: An Archaeology of Media Archaeology," 14. Yet Caroline Bassett shows that while "insects and insect behaviours ... are variously introduced as the subjects of specific chapters, they appear fleetingly, and only as they emerge through the work of philosophers or media theorists, where—despite the claims for material translation—they have become somewhat abstracted and largely theoretical figures" (Caroline Bassett, "I Am the Fly," *Radical Philosophy* 173 [May/June 2012]: 53). Once again the materials have gone missing, for all materiality is supposed to be media archaeology's very subject of study. Far from theorizing "the material," the main concern of media archaeology here too appears to be with the language and writing of other theorists and philosophers.

74. Ingold, *Being Alive*, 28.

75. The reference here is to Jussi Parikka, "Operative Media Archaeology: Wolfgang Ernst's Materialist Media Diagrammatics," *Theory Culture and Society* 28 (September 2011): 52–74.

76. Johanna Drucker, "Performative Materiality and Theoretical Approaches to Interface," *Digital Humanities Quarterly* 7, no. 1 (2013), http://www.digitalhumanities.org/dhq/vol/7/1/000143/000143.html#N100FD.

77. Nor is Drucker able to avoid repeating certain reductive refrains about critical theory herself. In particular, she reproduces the journalistic cliché that structuralism was "relativist": that "structuralist moves undermined the idea of intrinsic value, replacing it with relative value in anthropology, economics, linguistics, and cultural studies." What structuralism actually showed is that the relation between signifier and signifier is not fixed or natural, but rather the result of an underlying sign system, of which language is one possible example among many. This already arbitrary and contingent relation was then positioned as unstable in structuralism's poststructuralist variant. All of this is quite different from structuralism being relativist and arguing that things have subjective value only relative to the perception of the individual.

Perhaps this slippage is the result of Drucker's desire to champion a "humanistic perspective" (see chapter 2). Certainly things for her often seem to come back to the human individual, as when she writes in "Performative Materiality and Theoretical Approaches to Interface" that "every person produces a work as an individual experience, according to their disposition and capacity." For Drucker's related defense of a quite traditional "good" humanities scholarship as the professional system that supports the legacy of humanistic thought and knowledge—which she defines as the "serious engagement with the cultural record and 'all that it means to be human'"—and provides it with cultural authority, see her "Pixel Dust: Illusions of Innovation in Scholarly Publishing," *Los Angeles Review of Books*, January 16, 2014, http://lareviewofbooks.org/essay/pixel-dust-illusions-innovation-scholarly-publishing/.

78. Drucker, "Understanding Media." To provide another brief example of immaterial minds being set up in a binary relation to the material world, there is also the insistence of Manuel DeLanda that "any materialist philosophy must take as its point of departure the existence of a material world that is independent of our *minds*" (DeLanda, "'Any Materialist Philosophy'"). DeLanda thus provides a clear illustration of Ingold's point about materials being regarded as "varieties of matter" in theories of material culture, "that is, of the physical constitution of the world as it is given quite independently of the presence of activity of its inhabitants." The properties of these materials are thus positioned as "properties of matter, and are in that sense opposed to the qualities that the mind imaginatively projects onto them" (Ingold, *Being Alive*, 30).

79. Matters are not greatly illuminated by the questions Drucker herself raises at the end of the account of media archaeology and its points of possible connection with the digital humanities she provides in "Understanding Media": "In network analysis, objects are constituted by the codependent and distributed condition of circulation, flow, exchange, and relations. Is this an essentialist or relational vision of identity? Either way, the dynamics of reading always intervene, the act of retrieval, the engagement with an object, file, image, text that makes it happen in each instance. This poses questions about how we understand information, how it is

constituted, used, conceived, made present in and part of actual systems, networks, and media. How do we distinguish materiality and media? Is the first a property of the second? If something is characterized as *a medium*, then it is reified and its singularity suggests a thing-ness, physicality, that leads to discussion of properties and capacities. If media is the between-ness that permits relations themselves to exist, then mediation has no need for terms like 'flow,' 'exchange,' or 'transmission'—all of which are grounded in a model of delivery. A fully relational model of media and mediation does not depend on exchange, but on a cycle of differentiation as relation, the constitutive between-ness of configurations."

For one intriguing attempt at answering some of the questions raised here concerning the relation of the material and performativity, see the agential realist and posthumanist materialist account of performativity that is provided by Barad, drawing on the work of Kirby among others. For Barad, just as "materiality itself is always already figured within a linguistic domain as its condition of possibility," so matter and the material do not exist separately from representation, signification, the linguistic and discourse, but are themselves performative. She thus rethinks and reworks the "key concepts (materiality and signification) and the relationship between them" to demonstrate a sophisticated understanding, not just of materiality, but of the material, of matter and of metaphysics, as well as "the role 'we' play in the intertwined practices of knowing and becoming" (Barad, "Posthumanist Performativity," 801, 812).

80. I am aware that Graham Harman and Timothy Morton have both strived to distinguish materialism from the realism of object-oriented philosophy. See, for one example, Timothy Morton, *Hyperobjects: Philosophy and Ecology after the End of the World* (Minneapolis: University of Minnesota Press, 2013), 150.

81. As signing a book contract is considered by many to be an important act, legally and symbolically, using an expensive pen that is perhaps heavier and encased in enamel or metal, with a gold-plated nib that leaves a line varying in thickness, is often deemed to be more appropriate, conveying a better impression. If Braidotti goes along with this view and signed her name with a special fountain pen, then the ink was probably a fluid water-based dye. If she signed her name with a cheap plastic ballpoint, possibly because that is what comes most readily to hand in an era when a computer keyboard and mouse are frequently used to write, she was most likely doing so with an oil-based dye. Information on the material properties of these and other inks is provided at "Ink," *The Print Wiki: The Free Encyclopaedia of Print*, http://printwiki.org/Ink, accessed April 13, 2014. For some of the environmental issues associated with the production of ink, see "The Raw Material Report" in the September/October 2013 issue of *Ink World* magazine. Ink is a multibillion-dollar global industry, reportedly representing 60 percent of the cost of a printed page and being twice the price of the equivalent weight of Chanel No. 5.

82. For Braidotti, "matter-realists," among whose number she includes Barad, "combine the legacy of poststructuralist anti-humanism with the rejection of the classical

opposition 'materialism/idealism' to move towards 'Life' as a non-essentialist brand of contemporary vitalism and as a complex system" (*The Posthuman*, 158).

83. Ingold, *Being Alive*, 32.

84. Ibid., 28.

85. Tim Ingold, "When ANT Meets SPIDER: Social Theory for Arthropods," in *Being Alive*.

86. Braidotti, *The Posthuman*, 165. Further references are contained in the main body of the text.

87. As the petition initiated by Emily Kenway for Amazon to offer a living wage makes clear, the treatment of the workers in its warehouses by the "everything store" includes:

- A sack-if-you're-sick policy that sees you turfed out if you take three sick breaks in a three-month period
- Giving workers 15-minute breaks that start wherever they are in the giant warehouses
- 10-hour days
- Compulsory overtime
- Monitoring and timing toilet breaks
- Half-a-point if you're one minute late or more (three points and you're out)
- Paying the minimum wage or just above it, when it could well afford to pay the living wage
- A "performance console that tracks and logs workers' activities so they can be released if their `pick rate'" is too slow

(Emily Kenway, "@AmazonUK: Deliver the Living Wage in 2014," change.org, accessed March 1, 2014, https://www.change.org/en-GB/petitions/amazonuk-this-christmas-pay-the-living-wage-across-uk-operations?utm_source=action_alert&utm_medium=email&utm_campaign=44162&alert_id=ReBfwxLmGS_pEDpEuwclb)

88. As Ted Byfield points out, in many ways,

what's happened to publishing is like a case study in actor network theory. Lots of disparate factors intertwingling in ways that had a catastrophic effect on the industry. The Thor Power Tool Company Supreme Court ruling forced many publishers—hence everyone they touched from vendors to distributors and bookstores—toward a much higher-velocity model. The globalization of related businesses like paper production and printing had the usual effects [of] offshoring, with disparate effects: increased adoption of color printing, more of a "gambling" style because shipping delays became a bigger factor in responding to market successes, etc. And then there are the other phenomena we all know but are hard to pin down: proto-DIY movements (from mimeographed newsletters like YIPL to fanzines broadening out from music to [where DTP comes in] "style"); the intensification of licensing and merchandising across media from the '70s on; all the M&A nonsense of the '80s/'90s; and even, yes, changes in telephony—for example, the rapid fall in the cost of long-distance calls. All of these things, and many more, helped to drive different aspects of how publishing changed. (Ted Byfield, "Means of Production: The Factory-floor Knowledge," *nettime*, March 25, 2013, http://nettime.org/Lists-Archives/nettime-l-1303/msg00028.html)

89. Wolfe, *What Is Posthumanism?*, 117. For Wolfe, for example,

posthumanism can be defined quite specifically as the necessity for any discourse or critical procedure to take account of the constitutive (*and* constitutively paradoxical) nature of its own distinctions, forms, and procedures—and take account of them in ways that may be distinguished from the reflection and introspection associated with the critical subject of humanism. The "post-" of posthumanism thus marks the space in which the one using those distinctions and forms is not the one who can reflect on their latencies and blind spots while at the same time deploying them. That can only be done ... by another observer, using a different set of distinctions—and that observer, within the general economy of autopoiesis and iterability, need not be human (indeed, from this vantage, never was "human." (122)

90. To provide an example, Braidotti sees the relation between both the human and animal (79–80), and "nature-culture" (82), as a "continuum" (95). The "point about posthuman relations," she writes, "is to see the inter-*relation* human/animal as constitutive of the identity of *each*. It is a transformative or symbiotic relation that hybridizes and alters the 'nature' of each one and foregrounds the middle grounds of their interaction" (79). Kirby, however, builds on Barad's work to show how Derrida takes issue with the notion of such a continuum, and in fact considers it highly misguided. It is worth quoting Kirby at length on this point:

And here we conjure with the quantum resonance of Derrida's insights that refuse to supplement identity, and instead, open the text, or any individual identification, to an interiority whose articulating energy is the entire system. This expansive sense of interiority is the most difficult to think because thinking presumes cuts and divisions of simple separation, whereas these are, as Karen Barad explains it, ontoepistemological entanglements. ... Attesting to the difficulty, Derrida expresses his impatience with explaining the spatial and temporal complications in his arguments in terms of a continuum, an interpretation that for him is "worse than sleepwalking" and "scatterbrained." ... The scene of writing and its generality is not a "field" that is appropriately enormous because it must comprehend and include everything. The real paradox here that refigures the sense of quantum scale (the preconception that entanglement is only operative at the micro-level), is that there is no "everything" that preexists the relationality that *is* the scene of writing, the scene of ontological genesis as enfolding. For this reason the "entire scene" is already rehearsed and actively present in any and every "atom" of its instantiation/individuation. (Kirby, *Quantum* Anthropologies, xi)

91. Barad, "Posthumanist Performativity," 815; Sarah Kember and Joanna Zylinska, *Life after New Media: Mediation as a Vital Process* (Cambridge, MA: MIT Press, 2012). For Barad, "It is through specific agential intra-actions that the boundaries and properties of the 'components' of phenomena become determinate and that particular embodied concepts become meaningful. A specific intra-action (involving a specific material configuration of the 'apparatus of observation') enacts an *agential cut* (in contrast to the Cartesian cut—an inherent distinction—between subject and object) effecting a separation between 'subject' and 'object.' ... In other words, relata do not pre-exist relations; rather, relata-within-phenomena emerge through specific intra-actions" (Karen Barad, "Posthumanist Performativity," 815). In their own take on the term in *Life after New Media*, the "cut" thus functions for Kember and Zylinska as an "intrinsic component" of any critical and creative practice (xiv).

It is interesting in the context of the discussion of critique in chapter 1 that the word *critic* is derived from the Greek *kritikos*, "able to make judgments," and comes from *krinein*, meaning "to separate, decide," which has its origins in the

Proto-Indo-European word *skeri*, to cut. "Critic," *Online Etymology Dictionary*, accessed March 26, 2014, http://www.etymonline.com/index.php?term=critic &allowed_in_frame=0.

92. For Derrida, such "a double bind cannot be assumed" by definition; "one can only endure it in a *passion.*" What is more, it "cannot be fully analyzed: one can only unbind one of its knots by pulling on the other to make it tighter" (Jacques Derrida, *Resistances of Psychoanalysis* [Stanford, CA: Stanford University Press, 1998], 36).

93. Cary Wolfe, *What Is Posthumanism?*, 120.

94. Braidotti, *The Posthuman*, 87. Further references are contained in the main body of the text.

95. And not just of the humanities but also, as we have seen, of a "posthumanist posthumanism" that insists subjectivity is distributed "across species lines" and that we are bound to "nonhuman being in general, and within that to nonhuman animals, as the very condition of possibility for what we know and for sharing it with another" (Wolfe, *What Is Posthumanism?*, 125, 126).

96. James Leach, "'Step Inside: Knowledge Freely Available': The Politics of (Making) Knowledge Objects," in *The Politics of Knowledge*, ed. Patrick Baert and Fernando Dominguez Rubio (London: Routledge, 2010), 84. Further references are contained in the main body of the text.

97. These are not the only ways of creating trust; there are others. Liquid texts are much harder to pirate in the anticopyright sense than fixed and frozen ones, for example.

98. Janneke Adema, "Knowledge Production beyond the Book: Performing the Scholarly Monograph in Contemporary Digital Culture" (PhD diss., Coventry University, 2015); Adrian Johns, *The Nature of the Book: Print and Knowledge in the Making* (Chicago: Chicago University Press, 1998).

99. James Leach, "'Step Inside,'" 83.

100. See Pauline van Mourik Broekman, Gary Hall, Ted Byfield, Shaun Hides, and Simon Worthington., *Open Education: A Study in Disruption* (New York: Rowman & Littlefield International, 2014). This is very much contrary to Robin Osborne, for instance, who argues that "there can be no such thing as free access to academic research. Academic research is not something to which free access is possible. Academic research is a process—a process which universities teach (at a fee)" (Robin Osborne, "Why Open Access Makes No Sense," in *Debating Open Access*, ed. Vincent and Wickham, 97).

101. I do not want to simply let Braidotti—or indeed myself, given my own publication of *Pirate Philosophy* with a brand-name university press, as I say—off the hook

here. There are of course other possible reasons that theorists might make such (non)decisions. They include what Mark Fisher calls "capitalist realism." I am thinking in particular of the way in which the notion that "our 'inner beliefs"' are more important than those we publicly profess is vital to capitalist realism. As Fisher points out, we can "have left-wing convictions, and a left-wing self-image," so long as neither of these impact on our work "in any significant way!" (Mark Fisher, in Mark Fisher and Jeremy Gilbert, "Capitalist Realism and Neoliberal Hegemony: A Dialogue," *New Formations* 80 [2013]: 91, http://www.lwbooks.co.uk/journals/newformations/pdfs/80_fishergilbert.pdf).

102. To provide another example: discussing how the women's rights, antiracism, and pro-environment movements express, in the language of her nomadic theory, "both the crisis of the majority and the patterns of becoming of the minorities," Braidotti acknowledges that the task for critical theory is "to tell the difference between these different flows of mutation" (37–38). Are they a "symptom of the crisis of the subject, and for conservatives even its 'cause'"? Or are they actually an "expression of positive, pro-active alternatives" (37)?

103. http://aaaaarg.fail/; http://monoskop.org; http://libgen.org. For more on so-called piracy, as well as radical or guerrilla approaches to open access, see chapter 5; Adema and Hall, "The Political Nature of the Book"; Aaron Swartz, "Guerrilla Open Access Manifesto" (July 2008), http://archive.org/stream/GuerillaOpenAccessManifesto/Goamjuly2008#page/n1/mode/2up; and Aaron Swartz, "The Open Access Guerrilla Cookbook," January 13, 2013, http://pastebin.com/3i9JRJEA.

104. "Etymology of Pirate," *English Words of (Unexpected) Greek Origin,* http://ewonago.wordpress.com/2009/02/18/etymology-of-pirate/, posted by Johannes, February 18, 2009, accessed May 22, 2015.

105. The Piracy Project is an international publishing and exhibition undertaking run by the artists Eva Weinmayr, Lynn Harris, and Andrea Franke. It experiments with some of the philosophical, legal, and practical implications of book piracy through the creation of a platform designed to "explore the spectrum of copying, re-editing, translating, paraphrasing, imitating, re-organising, manipulating of already existing works" (http://www.andpublishing.org/projects/and-the-piracy-project, accessed December 9, 2014).

106. The emphasis on process here establishes a connection between the above definition of piracy of the ancient Greeks, which actually comes from maritime piracy, and the Living and Liquid Books series. For pirates operate on liquid, traditionally the sea. In fact this connection with the sea and fluidity may be one reason pirates are not to be treated as a legitimate human enemy. Daniel Heller-Roazen shows that to be counted within what the Roman philosopher Cicero terms the "immense fellowship of the human species" it may be that one is required to "belong to a community tied, like the Roman polity, to clearly delimited territory." In other words, one needs to live precisely "a sedentary life on land" (Daniel Heller-Roazen, *The*

Enemy of All: Pirates and the Law of Nations [Cambridge, MA: MIT Press, 2009], 17). If one does not do this, if one has a more fluid life—whether in terms of the sea, liquidity, or process—then one is at risk of being considered a pirate, this being one name for those whom we cannot necessarily treat as proper political adversaries. "For a pirate is not included in the number of lawful enemies, but is the common enemy of all" (Cicero, *On Duties* [Cambridge: Cambridge University Press, 1991], 141; Heller-Roazen, *The Enemy of All*, 16). In fact, according to the theory of monstrosity of the seventeenth-century philosopher Francis Bacon, as "the common enemy of human society" pirates are deserving of extermination (Francis Bacon, "An Advertisement Touching on Holy War" [1662], quoted in Peter Linebaugh and Marcus Rediker, *The Many-Headed Hydra: The Hidden History of the Revolutionary Atlantic* ([London: Verso, 2000], 62).

Of course, it is multinational corporations that today do not belong to a community tied to a clearly delimited territory and are stateless. Hence as we saw in chapters 1 and 3, some of them can use their statelessness to aggressively avoid paying taxes in the UK and have indeed been dubbed "pirate capitalists."

107. Just as the word *critique* is derived from the "Greek *kritike tekhne* 'the critical art'" ("Critique," *Online Etymology Dictionary*, accessed March 26, 2014, http://www.etymonline.com/index.php?term=critique; see chapter 1), so it can be argued that publishing and printing was, in its "pure state," originally an "Art." It was only later in its history that printing was incorporated by the Stationers Company (established as a corporate body in 1557), and its status reduced to that of a common "Mechanick Trade" dictated to by their "mercenary interests" (Johns, *The Nature of the Book*, 307). At the same time, that art can offer such a space—as indeed can science—serves to show that philosophy and theory do not have a monopoly on doing so as master-discourses.

108. Hélène Cixous in Frédéric-Yves Jeannet, and Hélène Cixous, *Encounters: Conversations on Life and Writing* (Cambridge: Polity, 2012), 65–66.

109. European Commission, "Open Access to Research Publications Reaching 'Tipping Point,'" European Commission Press release, Brussels, August 21, 2013, http://www.science-metrix.com/pdf/SM_EC_OA_Availability_2004-2011.pdf.

110. Goldsmith, *Uncreative Writing*, 14–15.

111. Ibid., 15.

112. Florian Cramer, "Post-Digital Writing," *Electronic Book Review* (December 2012), http://electronicbookreview.com/thread/electropoetics/postal; republished in Florian Cramer, *Anti-Media: Ephemera on Speculative Arts* (Amsterdam: Institute of Network Cultures, 2013).

113. John Lechte, "The Who and What of Writing in the Electronic Age," *Oxford Literary Review* 21 (1999): 140. See also Gregory L. Ulmer, *Teletheory: Grammatology in the Age of Video* (New York: Routledge, 1989).

114. For another more image-focused, less writing-centric example, see *Photomediations Machine,* curated by Joanna Zylinska and Ting Ting Chen (http://www.photomediationsmachine.net/). A sister project to the Open Humanities Press journal *Culture Machine, Photomediations Machine* adopts a process-based approach to image making by tracing the technological, biological, cultural, social, and political flows of mediation that produce photographic objects. As such, it constitutes a curated online space where the dynamic relations of mediation as performed in photography and other media can be encountered, experienced, and engaged in order to explore some of the forms theory and philosophy can take if they are thought and performed "with" media other than the written codex text.

Meanwhile, Janneke Adema provides a list of "Hybrid/Experimental (Scholarly) Books" on her Open Reflections blog: https://openreflections.wordpress.com/inventory-of-experiments, accessed June 7, 2015.

115. Goldsmith, *Uncreative Writing,* 26.

116. Sigi Jöttkandt, John Willinsky, and Shana Kimball, "The Role of Libraries in Emerging Models of Scholarly Communications," LIANZA, October 13, 2009, Christchurch, New Zealand, http://openhumanitiespress.org/Jottkandt_13-10-09_LIANZA.pdf.

117. OHP does not always use open source software. The Liquid Books series, for example, is not published on an open source platform but on PBworks, because the latter is easier to use for academics with little experience of editing wikis. Moreover, if we did always use open source software, that would be more of a program than an actual agential cut or decision. That said, we have made the decision that all of OHP's journals should be open source.

118. Most of OHP's funding comes indirectly: from publicly funded institutions paying our salaries as academics, librarians, technologists, and so forth (although not everyone who is part of OHP works for a university). We are simply using some of the time we are given to conduct research to create open access publishing opportunities for others. What is more, as Sigi Jöttkandt points out, "this largely volunteer effort is the norm rather than the exception" when it comes to no-fee journal publishing in many humanities fields, "in both OA and non-OA sectors" (http://openhumanitiespress.org/Jottkandt-Berlin5.pdf). Of course some academics may be fortunate enough to be given reduced teaching or administrative loads by their institutions for establishing and running publishing projects such as this. Others may have PhD students or graduate assistants they can ask to help with some of the work. Still others may be given an assistant paid for by their academic institution to help with the editorial labor. Another indirect source of funding occurs from institutions on occasion paying for the hosting of content. (My thanks to Marta Brunner for this last point.) Nevertheless, operating on an "academic gift economy" basis can actually be a significant source of strength to many independent humanities publishers. For one thing, it makes it easier to publish highly specialized, experimental,

inter- or transdisciplinary research. In other words, it supports research that, in challenging established disciplines, styles, and frameworks, may fall between the different stools represented by the various academic departments, learned societies, scholarly associations, and research councils and does not always fit into the kind of neat disciplinary categories and divisions with which for-profit publishers tend to order their lists but may nevertheless help to push a field in exciting new directions and generate important new areas of inquiry. Yet the gift economy can also be a potential source of weakness. It opens up many such initiatives to being positioned as functioning on an amateur, shoestring basis, almost as cottage industries. Compared to a series or list produced by a large, for-profit, corporate-owned legacy press, open access presses that use the gift economy as their model are far more vulnerable to the accusation that they are unable to ensure high academic standards in terms of their production, editing, copyediting, proofing, and peer reviewing processes. They are also more vulnerable to the suspicion that they are incapable of maintaining consistently high academic standards in terms of the quality of their long-term sustainability, the marketing and distribution services they can offer, their ability to be picked up by prestige-endowing indexes, and all the other add-on features a legacy press can often provide, such as journal archiving and contents alerts. As I have noted elsewhere, while this also applies to independent print journals, it is especially the case with regard to online-only journals, the vast majority of which are "still considered too new and unfamiliar to have gained the level of institutional recognition required for them to be thought of as being 'established' and 'of known *quality*'" (Gary Hall, *Digitize This Book! The Politics of New Media, or Why We Need Open Access Now* [Minneapolis: University of Minnesota Press, 2008], 60).

As I emphasized at the beginning of this chapter, it is precisely this perception of open access in the humanities that OHP is designed to counter by directly addressing these issues. Its aim is to ensure that OA publishing, in certain areas of the humanities at least, meets "the levels of professionalism our peers expect from publications they associate with academic 'quality"' (Sigi Jöttkandt, "No-Fee OA Journals in the Humanities, Three Case Studies: A Presentation by Open Humanities Press," Berlin 5 Open Access Conference: From Practice to Impact: Consequences of Knowledge Dissemination, Padua, September 19–21, 2007, http://openhumanitiespress.org/Jottkandt-Berlin5.pdf).

119. Roberto Esposito, *Communitas: The Origin and Destiny of Community* (Stanford, CA: Stanford University Press, 2010). See chapter 1 for more.

120. Such an approach has the potential to transform not just open access but academic publishing more generally. For instance, most open access publishers still operate as if they are traditional, commercial operations, even when they are not. So they function as separate entities as if they are in a state of competition with one another. But actually there is no need for open access publishers to operate like this, as digital objects are nonrivalrous in the sense that I can give you a digital copy of

this text but still retain my copy. Open access publishers can operate in a similarly nonrivalrous manner and can be much more intermeshed with one another.

This is certainly how we are trying to operate with regard to OHP. Another related example is provided by Project 5 of the International Association for Visual Culture (http://iavc.org.uk/2013/future-publishing-visual-culture-in-the-age-of-possibility): the Future Publishing: Visual Culture in the Age of Possibility project, which contains contributions from Mark Little and Marq Smith, Katherine Behar, Kathleen Fitzpatrick, and me. Constituted by Marq Smith and Mark Little as a collaborative, horizontal, open access project on the possible futures of publishing, all the pieces that make up this project were published in March 2013 online and simultaneously across a number of distinct scholarly, creative, and critical research platforms: the College Art Association; OHP's open access journal *Culture Machine*; the Institute for Modern and Contemporary Culture (University of Westminster); the IAVC, the *journal of visual culture*'s satellite website; another of OHP's journals, *Vectors: Journal of Culture and Technology in a Dynamic Vernacular*; and the Modern Language Association.

Chapter 5

1. N. Katherine Hayles, *How We Became Posthuman: Virtual Bodies in Cybernetics, Literature, and Informatics* (Chicago: University of Chicago Press, 1999), 4.

2. Bernardo Gutiérrez, "It Is Not a Revolution, It Is a New Networked Renaissance," *Occupy Wall Street*, January 29, 2014, http://occupywallst.org/article/theory-thursday-it-not-revolution-it-new-networked/#.UunUtZqJ-sE.twitter.

3. http://thepublicschool.org; http://aaaaarg.fail/. For more on autonomous, self-organized learning communities, see Pauline van Mourik Broekman, Gary Hall, Ted Byfield, Shaun Hides, and Simon Worthington, *Open Education: A Study in Disruption* (New York: Rowman & Littlefield International, 2014).

4. http://ica.org.uk/28063/Talks/Radical-Publishing-What-Are-We-Struggling-For.html.

5. Walter Benjamin, "The Author as Producer," *Understanding Brecht* (London: Verso, 1998).

6. Stanley Fish, "The Digital Humanities and the Transcending of Mortality," *New York Times*, January 9, 2012, http://opinionator.blogs.nytimes.com/2012/01/09/the-digital-humanities-and-the-transcending-of-mortality/; emphasis added. For those who are tempted to rush to judgment over blogs on the basis of Fish's characterization, it is worth bearing in mind that not all scholars privilege the long form and the monumental over the provisional and the ephemeral. Staying with Walter Benjamin, for example, we find that the feuilleton sections that featured in the newspapers and magazines of the Weimar Republic had a crucial impact on his prose style.

Taking over the bottom third of a paper's pages, these feuilleton sections consisted for the most part of cultural criticism. However, as Howard Eiland and Michael W. Jennings make clear in their recent critical biography of Benjamin, they also included "gossip, fashion commentary, and a variety of short forms—aphorisms, epigrams, quick takes on cultural objects and issues," and so forth. In fact, Eiland and Jennings go so far as to insist that the resulting "*kleine Form* or 'little form'" quickly came to be regarded as "the primary mode of cultural commentary and criticism in the Weimar Republic" (Howard Eiland and Michael W. Jennings, *Walter Benjamin: A Critical Life* [Cambridge, MA: Harvard University Press, 2014], 258). They certainly see this form as being much better suited to Benjamin than that he himself referred to as "the pretentious, universal gesture of the book" (Walter Benjamin, *Selected Writings* [Cambridge, MA: Harvard University Press, 1996–2003], 1:444). Consider the fact that Benjamin's *One-Way Street* is made up of sixty short pieces, a good many of which made their initial appearance in a newspaper or magazine's feuilleton section. Eiland and Jennings show how Benjamin very much "privileges the fragment over the finished work ... improvisation over 'competence' ... and waste products and detritus over the carefully crafted" in this book (259).

7. I am aware of the distinction some draw between the digital as a type of information and the digital as a medium. In the context of a discussion of the digital humanities, for example, Florian Cramer makes the point that "'the digital' is not a medium, but a type of information; information made up of discrete units (such as numbers) instead of an analog continuum (such as waves). The medium—the carrier—itself is, strictly speaking, always analog: electricity, airwaves, magnetic platters, optical rays, paper." This distinction is important, Cramer maintains, so as not to confuse "'electronic' and 'paper' with 'digital' and 'analog.'" For him, "technically seen, the movable type printing press is not an analog, but a digital system in that all writing [*sic*] into discrete, countable (and thus computable) units" (Florian Cramer, "Re: Digital Humanities Manifesto," *nettime*, January 22, 2009, http://www.mail-archive.com/nettime-l@kein.org/msg01331.html).

In this respect when I am using *digital* in this book, I am drawing largely on the more commonplace use of the concept to refer to those carriers associated with the "new media," a use that does indeed often confuse "'electronic' and 'paper' with 'digital' and 'analog,'" at least in Cramer's terms. But I am also conscious of the slightly different definition of the analogue and digital provided a few years before Cramer's intervention by another theorist very much concerned with the digital as a material form. In *What's the Matter with the Internet*, Mark Poster reveals how the

> term *analogue* refers to an aspect of the relation between a copy and an original. A taped recording of a sound, for example, transforms waves/cycles of air emitted by a person, for instance, into a configuration of metal oxide particles on a Mylar band. ... The relation between the configuration of the particles on the tape recording to the original waves/cycles of air is one of analogy; that is, the specific density and distribution of particles resembles the characteristics of the waves/cycles in their amplitude and frequency, their loudness and pitch. ...

Not so with digital reproduction. In this case the sound as waves/cycles is sampled some forty thousand times a second. ... The computer changes the input into a series of zeros and ones according to a formula that maps the sound event, both in loudness and pitch. The formula relating the characteristics of the sound to specific combinations of zeros and ones is arbitrary. In the case of digital recording there exists no resemblance, no analogy between the configuration of digits and the sound. (Mark Poster, *What's the Matter with the Internet* [Minneapolis: Minnesota University Press, 2001], 79)

For Poster, as for Cramer, print is not analogue but digital:

Print relies upon the alphabet, and alphabets are not analogue types of reproduction. Though early alphabets like ideograms are indeed analogue in that they depict in traces what they refer to, the Greek alphabet is composed of units that, in their combination, bear no relation to the meaning of the words they generate. The word *tree* does not look like a tree. Alphabets in this sense are digital in the sense in which I am using the term. ...

Digitization introduces yet another level of articulation of language, however, by introducing sequences of ones and zeros as representations of letters ... by introducing this change to zeros and ones, the material form of language can shift to the microworld world of electrons. In Katherine Hayles's words, "When a computer reads and writes machine language, it operates directly on binary code, the ones and zeros that correspond to positive and negative magnetic polarities." The basic difference introduced by the digital code is that it is translatable into a simple presence or absence and therefore into a minimal physical trace such as a pulse or electron. (Poster, *What's the Matter*, 81–82)

(This introduces us to another sense in which it can be said that there are no digital humanities—see chapter 2—as the humanities on this account have been digital since at least the invention of the alphabet.)

So we can see that, contrary to Cramer, the medium—the carrier—itself, although still "governed by its material determinations" (Poster, 82), is, strictly speaking, not always analog, in that the copy here does not always resemble or bear an analogous relation to the original, the carrier to the information. In the case of traces such as the pulses or electrons of electric language, this relation is arbitrary in a way that it is not for particles on a tape or grooves on a vinyl record. Indeed, I would want to interrogate the "political significance" Cramer attaches to the distinction he makes between the digital as a type of information and the digital as a medium, critically. This distinction is important for him because it reminds us "of the concrete materiality of the Internet and computing that involves the exploitation of energy, natural resources and human labor, as opposed to falsely buying, by the virtue of abstraction, into the 'immateriality' of 'digital media'" (Cramer, "Re: Digital Humanities Manifesto"). I would want to make at least two points in relation to this. First, as Poster's example shows, using the "digital" in this more commonplace sense does not necessarily imply "buying ... into the 'immateriality' of 'digital media,'" although of course some have certainly taken it this way. One can use the digital to refer to traces such as the pulses or electrons of electric language without seeing the digital as immaterial. Second, while I am attracted to Cramer's rigorous critical undermining of any simple contrast between new and old media, digital and analogue, I would insist a similarly rigorous critical approach be taken with regard to ideas of "concrete materiality" and immateriality. For one attempt at something of this kind in relation to contemporary theories of new materialism, media archaeology and performative materiality, see the discussion of materiality in chapter 4.

8. For a more developed account of this reading of the digital humanities, see chapter 2. As far as any possible posthumanities is concerned, it is interesting, given the emphasis we see Rosi Braidotti place on the "self-organizing (or autopoietic) force of living matter" (*The Posthuman* [London: Polity, 2013], 3) in chapter 4, that Cary Wolfe draws on Niklas Luhmann to describe the "social system called 'education'" as an auto-poietic system. Wolfe, *What Is Posthumanism?* (Minneapolis: University of Minnesota Press, 2010), 111. Indeed, it is perhaps worth emphasizing that for both Luhmann and Wolfe, there are "nonliving autopoietic systems" (Wolfe, 323, n. 41) and that it is not something that only applies to "humans, or to consciousness, or even to biological or organic systems" (Wolfe, 119).

9. Fish, "The Digital Humanities." The "'big tent' that the digital humanities can be," Fitzpatrick writes, is "a nexus of fields within which scholars use computing technologies to investigate the kinds of questions that are traditional to the humanities, or, as is more true of my own work, who ask traditional kinds of humanities-oriented questions about computing technologies" (Kathleen Fitzpatrick, "Reporting from the Digital Humanities 2010 Conference," *Chronicle of Higher Education*, July 13, 2010, http://chronicle.com/blogPost/Reporting-from-the-Digital/25473/).

10. Fish, "The Digital Humanities."

11. Kathleen Fitzpatrick, *Planned Obsolescence: Publishing, Technology, and the Future of the Academy* (New York: New York University Press, 2009, published in print in 2011), http://mediacommons.futureofthebook.org/mcpress/plannedobsolescence.

12. Fish, "The Digital Humanities."

13. Fitzpatrick, *Planned Obsolescence.*

14. Kathleen Fitzpatrick, January 10, 2012; response to Fish, "The Digital Humanities."

15. Fitzpatrick, *Planned Obsolescence.* As Fitzpatrick herself acknowledges, "This is something I wrestled with in revising *Planned Obsolescence*—really trying to think through how to represent a collaborative process within a text that ultimately attributes its authorship to me. I wound up quoting from and footnoting a lot of the conversations on the site, trying to be scrupulous about who led me to which idea, and how" (Kathleen Fitzpatrick, September 3, 2010; in response to Ted Striphas, "Performing Scholarly Communication," *Differences and Repetitions*, August 25, 2010, http://www.diffandrep.org/wiki/?q=node/11#footnoteref24_9inah3z). Moreover, Fitzpatrick's experiment with open peer review was designed to eventually appear as a conventional hard-copy, printed book, a version of *Planned Obsolescence* being published by New York University Press in 2011. So this project is still highly

papercentric. Is this one reason it is among the most often cited of its kind: because it is legitimated by its appearance in conventional codex book form and association with an esteemed legacy press?

16. Fitzpatrick, *Planned Obsolescence*.

17. Felix Stalder, "Autonomy and Control in the Era of Post-Privacy," *nettime*, July 4, 2010, http://permalink.gmane.org/gmane.culture.internet.nettime/4848.

18. Beverley Skeggs, "The Making of Class and Gender through Visualizing Moral Subject Formations," *Sociology* 39 (2005): 968. For more on academic subjectivation, see chapter 3.

19. Franco "Bifo" Berardi, *Precarious Rhapsody: Semiocapitalism and the Pathologies of the Post-Alpha Generation* (London: Minor Compositions, 2009), 127; Mario Tronti, "The Strategy of Refusal," Libcom.org, http://libcom.org/library/strategy-refusal-mario-tronti. Elsewhere, Bifo states that the "refusal of work … is better defined as a refusal of the alienation and exploitation of living time" (Franco "Bifo" Berardi, "Cognitarian Subjectivication," in *Are You Working Too Much? Post-Fordism, Precarity, and the Labor of Art*, ed. Julieta Aranda, Brian Kuan-Wood, and Anton Vidokle [Berlin: Sternberg Press, 2011], 138).

20. Peter Suber, "Watch Where You Donate Your Time," *Peter Suber*, January 7, 2012, https://plus.google.com/u/0/109377556796183035206/posts/QYAH1jSJG6L#109377556796183035206/posts; Gary Hall, "Withdrawal of Labour from Publishers in Favour of the US Research Works Act," *Media Gifts*, January 16, 2012, http://garyhall.squarespace.com/journal/2012/1/16/withdrawal-of-labour-from-publishers-in-favour-of-the-us-res.html.

21. At this point, I can appreciate that *Radical Philosophy*, the journal in which an earlier version of this chapter was first published, might have concerns about its own business model. If *Radical Philosophy* was made available on an open access basis in its entirety, would sales of annual subscriptions, paper copies, and individual PDFs from its online archive not fall dramatically? Would there no longer be sufficient funds to pay for running the journal as a result? There are a number of ways of responding creatively to this challenge, although they might involve making major changes to the nature and character of the journal. Even if they do not wish to move to publishing on an open access online-only basis using, say, Open Journal Systems, which would require less money by reducing the cost of production while enabling far greater and faster distribution, it is hard to see how the *Radical Philosophy* editorial collective can justify not moving to the delayed open access embargo model. Here, issues would be made available online open access after a certain period—usually between six and eighteen months—as academic journals rarely sell very many copies after the first year. Little income would therefore be lost because most universities and libraries would still take out subscriptions in order to have the most recent and up-to-date issue of *Radical Philosophy*.

That said, developments subsequent to the publication of that initial version of "Pirate Radical Philosophy" appear to have taken the decision out of the hands of the *Radical Philosophy* editorial collective. On March 28, 2014, the Higher Education Funding Council for England (HEFCE), together with the Arts and Humanities Research Council (AHRC) and Economic and Social Research Council (ESRC), published details of their new open access policy for research assessments after the 2014 Research Excellence Framework. To be eligible for assessment (and funding), this policy requires all peer-reviewed journal articles and conference proceedings (but not monographs and other long-form texts) accepted for publication after April 1, 2016, to "be deposited in an institutional or subject repository on acceptance for publication. The title and author of these deposits, and other descriptive information, must be discoverable straight away by anyone with a search engine. The manuscripts must then be accessible for anyone to read and download once any embargo period has elapsed" (HEFCE, "New Policy for Open Access in the Post-2014 Research Excellence Framework," March 28, 2014, http://www.hefce.ac.uk/news/newsarchive/2014/Name,94013,en.html).

22. See Peter Suber's Open Access Overview, http://www.earlham.edu/~peters/fos/overview.htm, and "Timeline of the Open Access Movement," http://www.earlham.edu/~peters/fos/timeline.htm.

23. For just one example of an interesting peer-to-peer publishing system, see Public Library: Memory of the World, http://www.memoryoftheworld.org/public-library/. For more on the possibility that this system if it "were to be successful could ... wind up being as bad for books as Amazon has been for small publishers and independent book shops," see David Garcia, "Books as 'a Gateway Drug'—In Discussion with Marcel Mars," *New Tactical Research*, February 14, 2014, http://new-tactical-research.co.uk/blog/1012/. A directory of commons-based peer production projects is available at http://directory.p2pvalue.eu. For a project showing the spread of peer-to-peer file sharing, see Nicolas Maigret and Brendan Howell, "The Pirate Cinema, 2012–2013," http://vimeo.com/67518774.

24. It is possible that this may even be one of the reasons for its relative success. As Gabriella Coleman writes with regard to the similar success of F/OSS, "In an era when identification with Right or Left, conservative or liberal, often functions as a politically paralyzing form of ideological imprisonment, F/OSS has been able to successfully avoid such polarization and thus ghettoization" (E. Gabriella Coleman, *Coding Freedom: The Ethics and Aesthetics of Hacking* [Princeton, NJ: Princeton University Press, 2013], 185–186).

25. John Willinsky, "Altering the Material Conditions of Access to the Humanities," in *Deconstructing Derrida*, ed. Peter Pericles Trifonas and Michael A. Peters (Hampshire: Palgrave Macmillan, 2005), 121, 123; Jacques Derrida, "The Future of the Profession or the University without Condition (Thanks to the "Humanities," What

Could Take Place Tomorrow)," in *Jacques Derrida: A Critical Reader*, ed. Tom Cohen (Cambridge: Cambridge University Press, 2001), 26.

26. Jacques Derrida, "From Restricted to General Economy: A Hegelianism without Reserve," *Writing and Difference* (London: Routledge & Kegan Paul, 1978), 260.

27. What I am attempting to interrogate here is the idea of the human author as creator of the original work. As chapter 1 shows, this idea underpins open access, Creative Commons, and free software. In view of this, I am not focusing too much on Creative Commons in this chapter. For a critical engagement with Creative Commons, see chapter 1.

28. The reading of copyright that follows is greatly indebted to my discussions with Cornelia Sollfrank and to the more detailed and subtle account of the relation between economic and moral rights she provides in an art world context in her "Performing the Paradoxes of Intellectual Property: An Artistic Investigation of the Increasingly Conflicting Relationship between Copyright and Art" (PhD thesis, University of Dundee, 2012).

29. Janneke Adema, "Scanners, Collectors and Aggregators: On the 'Underground Movement' of (Pirated) Theory Text Sharing," *Open Reflections*, September 20, 2009, http://openreflections.wordpress.com/2009/09/20/scanners-collectors-and-aggregators-on-the-%E2%80%98underground-movement%E2%80%99-of-pirated-theory-text-sharing/; Jeff Bewkes, quoted in Mark Sweney, "Top Piracy Target," *Guardian*, April 5, 2014, 9.

30. See, for example, http://fckvrso.wordpress.com/category/vro/ and https://magicmuscle.wordpress.com/2010/05/29/rip-aaaarg-org-and-fuck-verso/. In a 2010 interview, Aaaaarg's architect, Sean Dockray, states that it has received just two letters of protest in five years, the other being from Rem Koolhaas and OMA: Morgan Currie, "Small Is Beautiful: A Discussion with AAAARG Architect Sean Dockray,"*Masters of Media*, January 5, 2010, http://mastersofmedia.hum.uva.nl/2010/01/05/small-is-beautiful-a-discussion-with-aaaarg-architect-sean-dockray/).

31. "Lawrence & Wishart Statement on the Collected Works of Marx and Engels," http://www.lwbooks.co.uk/collected_works_statement.html, accessed April 26, 2014.

32. David Walters, quoted in David Kravets, "Capitalism Fells Communism in Marx-Engels Copyright Flap," *Ars Technica*, April 25, 2014, http://arstechnica.com/tech-policy/2014/04/capitalism-fells-communism-in-marx-engels-copyright-flap/. See also David Walters, "Response to Lawrence & Wishart statement on *MECW*," April 26, 2014, https://www.marxists.org/admin/legal/lw-response.html.

This is not to suggest that we should necessarily do away with the idea of the market altogether. Jacques Derrida contends that a distinction needs to be made between "a certain commercialist determination of the market," with its emphasis on "immediate monetaristic profitability," and a sense of the market as a "public space," which is actually a "condition of what is called democracy, the condition of the free expression of any and everyone about anything or anyone in the public space." Jacques Derrida, in Jacques Derrida and Bernard Stiegler, *Echographies of Television* (Cambridge: Polity, 2002), 47, 83, 44. As Michel Bauwens puts it, then: "I look forward to movements of social forces that can actually create these new forms of livelihood that are substantially outside of the old capitalist logic. I think we can legitimately make a difference ... between the market and capitalism. You can have market activity with a non-profit maximising utility: that is not capitalism. Yes, you are selling something, you are selling a service, but you are not accumulating capital. It is not the same thing. As long as we do not have infinite growth there is not a problem. The problem of capitalism, one of the problems, is infinite growth. ... That does not mean that a market is totally out of the question, we can have combinations, we can have a diverse economy with a commons logic with for-benefit institutions and with new market entities, which are not capitalist entities: that would be an alternative to the system we have today." Michel Bauwens in Sam Kinsley, "Towards Peer-to-Peer Alternatives: An interview with Michel Bauwens," *Culture Machine* 13 (2012): 15, http://www.culturemachine.net/index.php/cm/article/view/467/497).

33. Adrian Johns, "Piracy as a Business Force," *Culture Machine* 10 (2009): 45, 45, 51, 58 http://www.culturemachine.net/index.php/cm/article/view/345/348.

34. "Etymology of Pirate," *English Words of (Unexpected) Greek Origin,* http://ewonago.wordpress.com/2009/02/18/etymology-of-pirate/, posted by Johannes, February 18, 2009, accessed May 22, 2015. Interestingly, Stefan Zweig describes the philosopher Friedrich Nietzsche in similar terms, very much as a pirate (Stefan Zweig (1925), *Nietzsche* (London: Hesperus Press, 2013), 33–35.

35. Adrian Johns, in Serena Golden, "Piracy," *Inside Higher Ed,* February 3, 2010, http://www.insidehighered.com/news/2010/02/03/johns; and Adrian Johns, *Piracy: The Intellectual Property Wars from Gutenberg to Gates* (Chicago: University of Chicago Press, 2010).

36. Aaron Swartz (named author), "Guerrilla Open Access Manifesto," July 2008, http://archive.org/stream/GuerillaOpenAccessManifesto/Goamjuly2008#page/n1/mode/2up; Aaron Swartz, "The Open Access Guerrilla Cookbook," January 13, 2013, http://pastebin.com/3i9JRJEA.

37. Lawrence Lessig, *Free Culture: The Nature and Future of Creativity* (New York: Penguin, 2004), 53, 55.

38. "So Obama has thrown in his lot with Silicon Valley paymasters who threaten all software creators with piracy, plain thievery. ... Piracy leader is Google who streams movies free, sells advts around them. No wonder pouring millions into lobbying" (Rupert Murdoch, Twitter; cited in "Rupert Murdoch in Twitter Attack on

Google over 'Piracy,'" *Telegraph*, January 16, 2012, http://www.telegraph.co.uk/technology/news/9016762/Rupert-Murdoch-in-Twitter-attack-on-Google-over-piracy.html).

39. See Geoffrey Bennington, "Postal Politics," in *Nation and Narration*, ed. Homi K. Bhabha (London: Routledge, 1990), 131–132, on which the reading of the legislator in Rousseau that follows is based. Johns sees such ambiguities between the legitimate experimenter and the illegitimate thief or cheat as the result of the ineffectiveness of the way piracy has been policed in the past:

> Take the attempts of broadcasters to crack down on pirate radio *listeners* in the UK in the 1920s. Along with those who simply refused to buy a licence, many claimed to be using their radios for the purposes of scientific research, which entitled them to far cheaper "experimentor's licenses." ...
>
> This led to hopeless attempts to distinguish between legitimate "experimenters" and mere cheapskates. One scientific expert helpfully set out the distinction as follows: "The experimenter may listen to The Beggar's Opera purely for the purposes of comparison, but he must not listen to it for the purposes of enjoyment." (Adrian Johns, quoted in Matthew Reisz, "What's Mine Is Not Yours ... ," *Times Higher Education*, February 11, 2010, 51)

However, I would see this ambiguity as the result not so much of ineffectiveness on the part of lawmakers and those involved with policing intellectual property as the aporetic nature of such claims to authority, legitimacy, and legality.

40. This relates to one of the questions I would raise for Johns, whose work on piracy I admire very much: As a historian, does he have to have already decided what piracy is in advance in order to provide a history of pirates in books such as *Piracy: The Intellectual Property Wars from Gutenberg to Gates* (Chicago: University of Chicago Press, 2010), and *Death of a Pirate: British Radio and the Making of the Information Age* (London: Norton, 2010)?

41. Nor does this ambivalence and aporetic structure of authority apply only to the neoliberal state's sanctioning of contemporary privateers—for example, those involved in the privatization of public resources such as the education system, health service, and welfare state—and attempts to control and limit alternative appropriations of the Commons, including those associated with the Pirate Bay, Megaupload, and even Somalia, by condemning them as instances of piracy. (According to Rose George, the "Oxford English Corpus has noted a fourfold use of the word 'pirate' since 2007, with *Somali* being the most common modifier." Harvard Business School even chose Somali piracy, with its average profit margin of 25 to 30 percent, as "best business model of the year" in 2010 [Rose George, *Deep Sea and Foreign Going: Inside Shipping, The Invisible Industry That Brings You 90% of Everything* (London: Portobello Books, 2013), 133, 150.]) A similar ambivalence and aporetic structure of authority underpinned the legal status of the historical privateer/pirate. After all, this was someone who was quite capable of attacking the property (e.g., ships) of the very same government that employed him or her to commit state-legitimated "crimes" of piracy against the property of others. As Saint Augustine

famously remarks in *The City of God*, "Justice being taken away, then, what are kingdoms but great robberies? ... Indeed, that was an apt and true reply which was given to Alexander the Great by a pirate who had been seized. For when that king had asked the man what he meant by keeping hostile possession of the sea, he answered with bold pride, What you mean by seizing the whole earth; but because I do it with a petty ship, I am called a robber, while you who does it with a great fleet are styled emperor" (Augustine, *The City of God against the Pagans*, ed. R.W. Dyson [Cambridge: Cambridge University Press, 1998], IV.4, 148).

This is not to suggest we cannot have laws or make judgments and decide some things are legal and others not—as is shown by Murdoch's News International, where a legal decision concerning its telephone hacking was actually reached. None of this is a nihilism or relativism. But it does enable us to better understand the aporetic and violent basis on which we do so, and so try to assume and endure this ambivalence of authority in our politico-institutional practices and decisions rather than simply repeating it and acting it out. It is about acknowledging the necessity of such decisions, interruptions, or cuts and thinking about how to make them well, or at least better, and with what authority and legitimacy.

42. Johns, "Piracy," *Inside Higher Ed*; Johns, *Piracy: The Intellectual Property Wars*, 15.

43. Attributed to the Chinese communist leader Zhou Enlai, some say mistakenly, others that he misheard and he thought the question was actually referring to the Chinese revolution. http://en.wikiquote.org/wiki/Zhou_Enlai.

44. Andrew Ross, *Nice Work If You Can Get It: Life and Labor in Precarious Times* (New York: New York University Press, 2009), 167.

Chapter 6

1. Jacques Derrida, "The Word Processor," in *Paper Machine* (Stanford: Stanford University Press, 2005), 25–26.

2. See, for example, The Unbound Book Conference, which was held at Amsterdam Central Library and the Royal Library in Den Haag, May 2011, and where version 1.0 of this chapter was presented; and also the Arts and Humanities Research Council Digital Transformations Project: The Book Unbound, at Stirling University, http://www.bookunbound.stir.ac.uk, accessed June 1, 2015. For a somewhat different example, see the crowd-funded book publisher Unbound, http://unbound.co.uk/books.

3. *Oxford Dictionaries Online*, Oxford University Press, 2011, http://www.oxforddictionaries.com/search?searchType=dictionary&isWritersAndEditors=true&searchUri=All&q=bound&_searchBtn=Search&contentVersion=WORLD, accessed April 21, 2011.

4. Ulises Carrión, "The New Art of Making Books" (1975), in *Book*, ed. J. Langdon (Birmingham: Eastside Projects, 2010), n.p.

5. Florian Cramer points out that this also applies to artists' books that draw attention to the binding in their form, even if they may be playing with that binding, such as when an artist's book is made up of a collection of papers gathered in a folder or envelope, as with Isidore Isou's *Le Grande Désordre* (Pantin, France: l'auteur, 1963). Other examples include an experimental author placing either the loose pages or chapters of his or her novel randomly inside a box, as in the case of Marc Saporta's 1962 *Composition No1* (London: Visual Editions, 2011) and B. S. Johnson's 1968 *The Unfortunates* (London: Picador, 1999), respectively. Florian Cramer, "Unbound Books: Bound ex Negativo" (paper presented at the Unbound Book Conference, Amsterdam Central Library and the Royal Library, Den Haag, May 19–21, 2011); see also Johanna Drucker, *The Century of Artists' Books* (New York: Granary Books, 2004), 126–127.

6. As Christian Ulrik Andersen and Søren Bro Pold make clear with regard to e-book readers, "the universe of software control" inhabited by the Amazon and Google, among others, "is a universe designed to ensure that only things that do not violate copyright and corporate control can happen. The copyrights, trademarks, patents etc. have all been wrapped up in contractual and restrictive end-user licence agreements—a change that has been described with the concept of 'controlled consumption.'" As we saw in chapters 2 and 3, the "licensing culture furthermore includes a thorough monitoring of readers and reading behaviour (e.g. the Amazon Whispernet)" (Christian Ulrik Andersen and Søren Bro Pold, "Post Digital Books and Disruptive Literary Machines," paper published on the site of Hold the Light: The 2014 ELO Conference, Milwaukee, WI, June 19–21, 2014, 3, 4, http://conference.eliterature.org/sites/default/files/papers/Andersen_Pold_v2_0.pdf).

For more on copyright and its relation to notions of authorship, originality, attribution, integrity, and disclosure, see chapter 5.

7. Mackenzie Wark, *A Hacker Manifesto* (Cambridge, MA: Harvard University Press, 2004).

8. McKenzie Wark, "Copyright, Copyleft, Copygift," *Open: Cahier on Art and the Public Domain* 12, Nai Publishers, SKOR (2007), 27.

9. Ibid., 26.

10. For some of the advantages of free "offline" access—what is termed Open Access Prime—see Peter Suber, "Free Offline Access: A Primer on OA (OA Prime)," *SPARC Open Access Newsletter*, no. 157, May 2, 2011, http://www.earlham.edu/~peters/fos/newsletter/05-02-11.htm.

11. *Self-Archiving FAQ*, written for the Budapest Open Access Initiative (BOAI): http://www.eprints.org/openaccess/self-faq/, accessed May 4, 2013. For a more

recent account of the legal situation in relation to the self-archiving of drafts and preprints, see Charles Oppenheim, "Guest Post: Charles Oppenheim On Who Owns the Rights to Scholarly Articles," *Open and Shut?* February 4, 2014, http://poynder.blogspot.co.uk/2014/02/guest-post-charles-oppenheim-on-who.html. Oppenheim's post was written in the context of the decision taken by Elsevier, owners of the academic social networking site Mendeley, to use the Digital Millennium Copyright Act to send take-down notices to rival social networking sites such as Academia.edu, along with a number of individual universities including the University of California, Irvine.

12. Ibid.

13. Wark, "Copyright, Copyleft, Copygift," 24.

14. http://epress.anu.edu.au/; http://www.aupress.ca/; http://mayflybooks.org/; http://www.openbookpublishers.com/.

15. Graham Harman, *Prince of Networks: Bruno Latour and Metaphysics* (Melbourne: re.press, 2009); John Carlos Rowe, *The Cultural Politics of the New American Studies* (Ann Arbor, MI: Open Humanities Press, 2012). A list of open access book publishers is available at http://oad.simmons.edu/oadwiki/Publishers_of_OA_books.

16. Ted Striphas, *The Late Age of Print: Everyday Book Culture from Consumerism to Control* (New York: Columbia University Press, 2009), http://www.thelateageofprint.org/download/; E. Gabriella Coleman, *Coding Freedom: The Ethics and Aesthetics of Hacking* (Princeton, NJ: Princeton University Press, 2013), http://codingfreedom.com/. For more, see Jill Walker Rettberg, "How I Published My Scholarly Book with an Open Access CC-BY License," *jill/txt*, October 19, 2014, http://jilltxt.net/?p=4117.

To be fair, Marcus Boon has persuaded Harvard University Press to make a copy of his 2013 book, *In Praise of Copying* (Cambridge, MA: Harvard University Press, 2013), freely available online as a PDF using a CC license: http://www.hup.harvard.edu/features/in-praise-of-copying/. So perhaps Wark's request for Harvard to do something similar with *A Hacker Manifesto* was just a little too ahead of the game.

17. That said, more and more publishers are slowly beginning to experiment with publishing books open access. A recent study reveals that 35 percent of scholarly publishers offer open access monographs, many on a gold, "author-pays" (and thus income-generating) model. However, "such works currently account for less than 5% of their book collections" ("Open Access Books Slowly on the Rise," booktrade.info, May 27, 2015, http://www.booktrade.info/index.php/showarticle/59740/nl#.VWbq_UevOOQ.twitter). No doubt this is due in large part to the high costs that are often involved, and the difficulty authors have in obtaining the necessary funding. By early 2014, for example, Palgrave Macmillan, which charges £11,000 to make a book available gold author-pays open access, had published only one title on this

basis in the area of Film, Culture and Media Studies (Felicity Plester, Palgrave Macmillan, speaking at the Challenges and Changes in Academic Publishing event, University of Sussex, March 5, 2014).

18. *Self-Archiving FAQ.*

19. http://usefulchem.wikispaces.com/.

20. Richard Poynder, "Interview with Jean-Claude Bradley: The Impact of Open Notebook Science," *Information Today*, September 2010, http://www.infotoday.com/IT/sep10/Poynder.shtml.

21. Gary Hall, "Open Humanities Notebook," *Media Gifts*: http://www.garyhall.info/journal/.

22. In many ways this is merely a variation on what is actually quite an old publishing practice. James Bridle notes that "in the Victorian era, book-first works weren't considered serious: you were a 'proper writer' if your work first received serial publication in a newspaper or magazine. That was the mark of editorial quality. As books became both more widely affordable and better produced, the focus shifted to hardbacks and paperbacks—and has remained there. Even paperbacks are often turned down for review: much hardback publication is still essentially in order to receive media notice to publicise the cheaper edition" (James Bridle, "An Everyday Tale of Print Pride and Prejudice," *Observer: The New Review*, November 11, 2012, 39).

23. In a 2012 interview, Gita Manaktala, MIT Press editorial director, acknowledges: "Exposure that reduces the audience for subsequent book sales may be a problem. Books are costly to produce whether in print or digital formats. The costs to develop, edit, typeset, design, produce, market, and distribute them are substantial and must be recovered through sales—usually within the first two years of the book's life. For a typical monograph, it is unclear whether and what kind of pre-publication exposure might help or hurt the publisher's ability to recover its investment in the work. Anecdotal evidence exists but large-scale studies are lacking. A serious study would control for the nature and length of pre-publication exposure as well as for the prices of paid editions and a range of other factors. Few (if any) university presses are in a position to undertake such large-scale experiments with their own publications." Gita Manaktala in Adeline Koh, "What Is the Future of Academic Publishing? An Interview with Gita Manaktala from MIT Press," *Chronicle of Higher Education*, July 17, 2012, http://chronicle.com/blogs/profhacker/what-is-the-future-of-academic-publishing-an-interview-with-gita-manaktala-from-mit-press/41335.

That said, any policy of rejecting books on the basis that a version of some of the material they contain is already available online is going to be increasingly difficult to sustain, certainly as far as academic authors in the UK are concerned. The Higher Education Funding Council for England's open access policy for research assessments after the 2014 Research Excellence Framework means that to be eligible for assessment (and funding), all peer-reviewed journal articles and conference proceedings (but not monographs and other long-form texts) accepted for publication after April 1, 2016, have to "be deposited in an institutional or subject repository on acceptance for publication" (HEFCE, "New Policy for Open Access in the Post-2014 Research Excellence Framework," March 28, 2014, http://www.hefce.ac.uk/news/newsarchive/2014/Name,94013,en.html).

24. See Ted Striphas, "Performing Scholarly Communication," *Differences and Repetitions,* August 25, 2010, http://wiki.diffandrep.org/performing-scholarly-communication.

25. For more on the Liquid Books series, see chapter 4. For the contents of *Media Gifts,* including chapter titles and summaries, see http://garyhall.squarespace.com/open-book/.

26. See Gary Hall and Clare Birchall, eds., *New Cultural Studies: The Liquid Theory Reader* (Ann Arbor, MI: Open Humanities Press, 2008, ongoing), http://liquidbooks.pbworks.com/New+Cultural+Studies:+The+Liquid+Theory+Reader.

27. Gary Hall, "Pirate Philosophy Version 2.0," available at the time of this writing at http://www.torrenthound.com/hash/94bfd0a095f6bc76d6c3862fdc550011d1702814/torrent-info/Pirate-Philosophy-2-0-doc. "Pirate Philosophy Version 1.0" appeared in *Culture Machine* 10 (2009), http://www.culturemachine.net/index.php/cm/article/view/344/426.

The way the project works is as follows. A text titled "Pirate Philosophy Version 1.0" was published as the opening essay to the tenth edition of the *Culture Machine* journal, which itself had the theme of pirate philosophy. However, this text was available there for a limited period only. After two months, it was placed on a torrent search engine and directory as "Pirate Philosophy Version 2.0" and the original deleted from the *Culture Machine* site. As soon as someone downloaded the torrented version—which occurred on the same day I made it available, May 25, 2009—the original file was destroyed. So from May 25, 2009, all copies of this text have been "pirate" copies. Originally placed on the Mininova torrent directory, "Pirate Philosophy Version 2.0" is also available at the time of this writing from Aaaaarg, Mininova and Kickasstorrents, among other places.

28. Tim Ingold, "When *ANT* Meets *SPIDER*: Social Theory for Anthropods," in *Being Alive: Essays on Movement, Knowledge and Description* (London: Routledge, 2011), 89–94. Although I find the theory of the meshwork of Ingold's SPIDER more convincing than the Latourian actor-network theory of his ANT, I have for the most part retained the language of networks here, not least because of its familiar association with computer networks, file-sharing networks, pirate networks, and so on.

29. In a variation on this theme, Tim McCormack adopts the term *multigraph.* This describes "a monograph reconceived as a gathering of many content components, structures, and pathways for creation and use" (Tim McCormack, "From Monograph to Multigraph: The Distributed Book," *Impact of the Social Sciences,* January 17, 2013,

http://blogs.lse.ac.uk/impactofsocialsciences/2013/01/17/from-monograph-to-multigraph-the-distributed-book/).

30. In *Protocol: How Control Exists after Decentralization* (Cambridge, MA: MIT Press, 2004), Alexander Galloway distinguishes between decentered and distributed networks as follows:

> A decentralized network ... has *multiple* central hosts, each with its own set of satellite nodes. A satellite node may have connectivity with one or more hosts, but not with other nodes. Communication generally travels unidirectionally within both central and decentralized networks: from the central trunks to the radial leaves.
>
> The distributed network is an entirely different matter. ... Each point in a distributed network is neither a central hub nor a satellite node—there are neither trunks nor leaves. (11)

Far from these two kinds of networks being opposed, however, I would suggest that *Media Gifts* is, in these senses, both decentered and distributed.

31. Maurice Blanchot, *The Book to Come* (1959; Stanford, CA: Stanford University Press, 2003), 234. "Gathered through Dispersion" is a subheading used in "A New Understanding of Literary Space," the second section of the chapter "The Book to Come," in Blanchot's book of the same name. As we have seen, for all that *Media Gifts* is not tightly bound, such diversity nevertheless has to be gatherable; otherwise it would not be capable of constituting a book. This is what Derrida refers to as the "insoluble" nature of Blanchot's tension: for how can "infinite diversity" be gathered? Jacques Derrida, "The Book to Come," in *Paper Machine* (Stanford, CA: Stanford University Press, 2005), 14.

I want to stress three points. First, it is important that any such print or online book version of *Media Gifts* is regarded as merely part of the constantly changing constellation of projects, texts, websites, archives, wikis, and other traces I have described. In other words, the book version should not be positioned as providing the overarching, final, definitive, most systematic, significant, or authentic version of any material that also appears in other iterations, forms, and places; nor should it be taken as designating a special or privileged means of understanding the media projects with which it is concerned. It is, rather, one knot or nodal point in this meshwork, one possible means of access to or engagement with it. There are others, including the Liquid Books and Pirate Philosophy Versions 1.0 and 2.0 projects I have referred to.

Second, I also want to draw attention to the way the *Media Gifts* project emphasizes the violence inherent in any such "cut" that publishing this material as a book represents—while at the same time acknowledging that this violence in inescapable since, as we have seen, a book has to be gathered and bound; otherwise it is not a book. An ethical response to *Media Gifts* would involve both actively making such cuts and working toward minimizing this violence.

Third, and adapting an idea of Kenneth Goldsmith, difficulty can thus be defined in relation to *Media Gifts* as much in terms of "quantity (too much to read)" as it is by "fragmentation (too shattered to read)"—thus perhaps moving "away from modernist notions of disjunction and deconstruction" somewhat (Kenneth Goldsmith, *Uncreative Writing: Managing Language in the Digital Age* [New York: Columbia University Press, 2011], 12).

32. http://anthologize.org.

33. Derrida, "The Book to Come," 5.

34. Graham Harman, "Quick Thoughts on What Might Happen," *Object-Oriented Philosophy*, July 29, 2009, http://doctorzamalek2.wordpress.com/2009/07/29/quick-thoughts-on-what-might-happen/.

35. The Article of the Future project is available at http://www.articleofthefuture.com/project, accessed December 9, 2014. For more on PLoS Hubs, see http://hubs.plos.org/web/biodiversity/about;jsessionid=97C4923247B71A5DA083B50CAB39F8FB.

36. This is contrary to the more traditional view that has prevailed to date. The latter is encapsulated by Stefan Collini as follows: "In the world of science and scholarship, repeating the same argument, re-using the same material, even re-publishing previously published work, all tend to be frowned upon as redundant and self-indulgent. A scholarly case once properly made and substantiated can then be consulted by anyone interested in the topic: there is no need to re-state it since the original remains both authoritative and accessible" (Stefan Collini, *What Are Universities For?* [London: Penguin, 2012], 118).

37. Derrida, "The Book to Come," 7–8.

38. http://www.amazon.co.uk/b?ie=UTF8&node=2445826031; http://www.ted.com/pages/tedbooks; http://www.palgrave.com/pivot/; http://litlab.stanford.edu/?page_id=255.

39. http://mediacommons.futureofthebook.org/tne/how-it-works, accessed June 1, 2015.

40. http://nanopub.org/wordpress.

41. Johanna Drucker, "Scholarly Publishing: Micro Units and the Macro Scale," *Amodern* 1, January 16, 2013, http://amodern.net/article/scholarly-publishing-micro-units-and-the-macro-scale/.

42. http://www.plos.org/cms/node/481; http://pressforward.org.

43. http://figshare.com; http://openarchaeologydata.metajnl.com.

44. http://www.campus-roar.ecs.soton.ac.uk—work on the CampusROAR project was completed in 2012; http://www.jisc.ac.uk/whatwedo/programmes/inf11/inf11scholcomm/larkinpress.aspx—my understanding is that the Larkin Press project is now no longer running. For a further example of a move in this direction, however, see the "Active Archives" project of the Brussels-based feminist collective Constant (http://www.constantvzw.org/site/-Active-Archives,110-.html, accessed

December 9, 2014). This is a research project (ongoing since 2006) devoted to the development of experimental online archives with the aim of "creating a free software platform to connect practices of library, media library, publications on paper (as magazines, books, catalogues), productions of audio-visual objects, events, workshops, discursive productions, etc. Practices which can take place on line or in various geographical places, and which can be at various stages of visibility for reasons of rights of access or for reasons of research and privacy conditions. ... regular workshops will be organised to stimulate dialog between future users, developers and cultural workers and researchers" (Constant, "Manifesto for an Active Archive," January 16, 2009, http://activearchives.org/wiki/Manifesto_for_an_Active_Archive).

45. As Derrida puts it in this case with regard to literary, poetic, and legal texts: "No critic, no translator, no teacher has, in principle, the right to touch the literary text once it is published, legitimated, and authorized by copyright: this is a sacred inheritance, even if it occurs in an atheistic and so-called secular milieu. You don't touch a poem! Or a legal text, and the law is sacred—like the social contract, says Rousseau" (Jacques Derrida, '"Others Are Secret Because They Are Other,"' in *Paper Machine*, 142).

46. http://www.culturemachine.net; http://openhumanitiespress.org; http://disruptivemedia.org.uk/portfolio/comc; http://disruptivemedia.org.uk.

47. Mark Amerika, "Sentences on Remixology 1.0," in *remixthebook* (Minneapolis: Minnesota University Press, 2011). See also http://www.remixthebook.com, the online hub for the digital remixes of many of the ideas and theories in Amerika's *remixthebook*.

48. Sol LeWitt, "Sentences on Conceptual Art," first published in *0–9*, New York (1969), and *Art-Language*, England, May (1969): http://www.altx.com/vizarts/conceptual.html.

49. See, for one example, "Future Books: A Wikipedia Model?" *Technology and Cultural Form: A Liquid Reader* (Ann Arbor, MI: Open Humanities Press, 2010), http://liquidbooks.pbworks.com/w/page/32057416/INTRODUCTION%20TO%20THE%20LIQUID%20READER, discussed in chapter 4.

50. James Leach captures some of the complexities of the situation when he writes: "If you are made up of—and manifest physically—other people's work, input, substance and knowledge, then you do not in fact own yourself or anything you produce as an individual. There is no project that is not already the project of other people as well, because they are part of you as a person. In fact, complex exchange systems that substitute persons for wealth show that there is nothing else to a person than their make-up in the work and thought of others. People, if you will, are the projects of other people. Knowledge in these places is similarly constituted. It does not come from any single creator, just as the person does not come from a single progenitor. Knowledge is part of what people are" (James Leach, "Creativity,

Subjectivity, and the Dynamic of Possessive Individualism," in *Creativity and Cultural Improvisation,* ed. T. Ingold and E. Hallam [Oxford: Berg, 2007], 112).

Even the title of this chapter and its topic were generated by others: Mark Amerika and also the organizers of The Unbound Book Conference, held at Amsterdam Central Library and the Royal Library in Den Haag in 2011 and where version 1.0 of this material was first presented.

This is not to suggest such "togetherness" is without difference and antagonism. There is not the space here to go through each of the projects featured in *Media Gifts* and detail the different kinds of authorship at play. Suffice it to say that some of the forms of explicitly multiple authorship I am referring to here on occasion do indeed manifest themselves as an expanded or enlarged authorship that works collaboratively to produce more or less agreed-on projects. However, they also include forms of multiple authorship that involve authors, groups, and actors developing different projects that are not agreed on and are in fact often in conflict with one other.

Nor should the relationship between explicitly multiple and single forms of authorship be seen in either-or terms. Depending on the particular persona, voice, or semiotic function employed, I—like both Leach and Amerika—at times put my name on the cover and spine of a print-on-paper codex book, even though I am aware it is written by the other in me, and that there are, as I say, multiple "I"s.

51. See http://unbindingthebook.com and http://theunbook.com. As an article in the first UK edition of *Wired* magazine has it, "The unbook is an open-source work in progress, inviting outside participants to contribute ideas, resulting in version 1.2, 1.33 and so on. Once you start thinking of the book as an output medium for the web, all sorts of things become possible—such as travel guides created and curated by groups of friends" (Russel M. Davies, "Why We're Logging on to the Papernet," *Wired,* May 2009, 44).

52. "Sacred Texts: Codex Sinaiticus," *British Library: Online Gallery,* accessed May 4, 2013, http://www.bl.uk/onlinegallery/sacredtexts/codexsinai.html.

53. For one suitably "unbound" (tele)visual account of the liquid, living nature of the Codex Sinaiticus, see "The Beauty of Books (BBC)—Ancient Bibles, the Codex Sinaiticus," *YouTube,* April 30, 2011, extracted from BBC, *The Beauty of Books,* Episode 1, Ancient Bibles, 2011. TV BBC 4, February 7, 2011, 20.30, http://www.youtube.com/watch?feature=player_embedded&v=kCkyakphoKE, accessed May 4, 2013.

54. Variations on this argument can also be developed around Homer's *Iliad* and *Odyssey* (we do not know who or how many people wrote them, or when), Dante's *Divine Comedy* (no original has survived and we do not know for certain what language it was written in), and Saussure's *Course in General Linguistics* (which was compiled from notes taken by his students).

55. Adrian Johns, *The Nature of the Book: Print and Knowledge in the Making* (Chicago: University of Chicago Press, 1998), 31.

56. In "The Book to Come" Derrida notes that before it meant "book," the Latin word *liber* originally designated the living part of the papyrus bark, and thus the paper, that was used as a support for writing (6). But it is not just books that are living. Tim Ingold shows how the very word *material* is derived from the Latin for "mother," *mater*, having a complex history that relates it through Latin and Greek to wood, "which is or has been alive." Rather than the "inanimate stuff typically envisioned by modern thought, materials in this original sense are the active constituents of a world-in-formation," he writes (Tim Ingold, *Being Alive: Essays on Movement, Knowledge and Description* [London: Routledge, 2011], 27–28, quoting Nicholas J. Allen, "The Category of Substance: A Maussian Theme Revisited," in *Marcel Mauss: A Centenary Tribute*, ed. Wendy James and Nicholas Allen [New York: Berghahn, 1998], 177).

57. Ben Fry provides an animated visualization of this state of affairs with regard to Darwin's *The Origin of the Species* in *Preservation of Selected Traces* (2009), http://benfry.com/traces, accessed December 9, 2014. For more contemporary examples of how even print-on-paper texts are not fixed, stable, reliable, or permanent, see The Piracy Project, "The Impermanent Book," *Rhizome*, April 19, 2012, http://rhizome.org/editorial/2012/apr/19/impermanent-book/.

58. Gary Hall, *Digitize This Book!* (Minneapolis: Minnesota University Press, 2008), 161.

Index